Physics of Community Course Notes for Fall 2015

Quantifying collective decisions in a world of power, dialogue, conflict, selfishness, corruption, altruism, and creativity.

Darrell Velegol

Physics of Community Course Notes for Fall 2015

Quantifying collective decisions in a world of power, dialogue, conflict, selfishness, corruption, altruism, and creativity.

Copyright © 2015 by Darrell Velegol
Wild Scholars Media

ISBN-13: 978-1515027171
ISBN-10: 1515027171

This book is written in 10 pt Times New Roman font.

Darrell Velegol
2015 October 31

Dedicated to my teachers ...

David A. Velegol, Sr., who demonstrated how theory joins practice.
Kay Bilal, who introduced me to the beauty of chemistry.
Joseph A. Shaeiwitz, who introduced me to chemical engineering.
Robert E. DiClerico, who lit the flame for "Physics of Community".
Herbert A. Simon, who inspired me to model the social sciences.
John L. Anderson, who instilled in me how to frame a model.
Jack Matson, who showed me that failure is essential for creativity.
Ted Alter, who taught the roles of power and dialogue in democracy.

Stephanie Velegol, who taught me about love ... and whose ideas and encouragement give breath to this venture.

Brief Contents

Detailed Contents

Preface

*There is really nothing more pathetic than to
have an economist or a retired engineer try
to force analogies between the concepts of
physics and the concepts of economics. –
Paul A. Samuelson*

My PhD thesis was born in a single night. After going to bed late one night, I slept for a bit, then woke up with an idea. I wrote it down. Then I returned to sleep, only to awaken again with another idea. I wrote it down. This happened all night. When I awoke in the morning, I had the outline for my entire PhD thesis in Chemical Engineering, which then simply required three years of hard work.

The ideas described in this book had a very different kind of gestation and birth. In my senior year at West Virginia University, I had the opportunity to take a one-on-one course in American Government with Professor Robert E. DiClerico. By the nature of the course, each week I read one book – say, on the founding of the democratic republic, or iron triangles – wrote a roughly ten page essay, then discussed the book and essay with Professor DiClerico for an hour. One week the book was Piven and Cloward's *Why Americans Don't Vote*. The authors wrote about the flux of voters to the polls and the many resistances of getting to those polls. Flux and resistance are two parts of Ohm's law of electricity, and it dawned on me that they never discussed the third part of Ohm's law: a driving force, analogous to voltage or pressure. So I framed my essay in terms of Ohm's law. Professor DiClerico gave me an A-on the essay (see the scans on the next page), but on the part where

I discussed Ohm's law, he bracketed the section and wrote "Nice". Since that Fall 1991 experience, I have been exploring how ideas from Chemical Engineering could be used to explain human decision making. I had fun doing this, even while I did my "day job" as a faculty member in Chemical Engineering at Penn State.

Several events catalyzed greater commitment in my study of human decision making, starting in January 2013 when my friend and colleague Jack Matson asked me to co-teach a Massive Open Online Course (MOOC) in "Creativity, Innovation, and Change". In Fall 2013 we had over 150,000 students enrolled, from over 190 nations around the world; our course is one of the top 10 largest MOOCs taught. Preparations took about 1000 hours that year, and it revealed to me the power of pouring an immense amount of time into a single subject, as well as the value creation that results from doing something quite unexpected.

Then on 2013 May 04, my family and I went to Independence Square in Philadelphia. I wanted to proclaim a "declaration of independence" from my former research path, toward one emphasizing "Physics of Community".

The next event happened in Fall 2014, when I convinced my department head to let me teach a Chemical Engineering technical elective CH E 497B: Physics of Community. I worked with 28 students that Fall, and increased my appreciation for both the power and limitations of fields like game theory, information theory, and network theory. Some students, especially Paul Suhey and Michael Davidson, I thank for working with me and discovering with me along the way. During the same semester, I sat in on a course CED 417: Power, Conflict, and Community Decision Making, taught by my friend and colleague Professor Ted Alter. We also started meeting for two hours per week, discussing the roles of power, dialogue, advantage, and disadvantage in collective decisions at Penn State and in democracy more generally.

Voting in America

The Do? Should?

and How? Issues

Darrell Velegol

6 November 1991

The Mathematics of Responsible Voting

We have two rate equations:

$$Turnout = \frac{driving\ force}{resistance}$$

$$Turnout_{responsible} = \frac{driving\ force_{responsible}}{resistance_{responsible}}$$

If resistance decreases, then turnout increases. When turnout increases, responsible voters would have an incentive to increase responsible voting, so they would increase the driving force for responsible voting. The incentive would be to maintain a favorable, responsible electorate with some predictability. Once this second driving force increases, responsible turnout would increase.

This book is thus the result of about 25 years of extensive notes, reading, deriving, calculating, and discussing. I provide what I believe is a new and rigorous way to analyze collective decisions, from a Chemical Engineering perspective. The perspective follows from – but differs from – the models of von Neumann or Shannon, revealing assumptions and offering new possibilities. But it leads to the following ansatz: Human decisions can be treated mathematically as sets of chemical reactions of metaphorical molecules – that is, "decision reactions" involving "knowlecules" – within a broader framework of information flows and separations in networks and dialogical processes represented by process flow diagrams. Concepts like Gibbsian games, decision reactions, perception functions, entropic choices, and pain potentials permeate the pages of this book, and if experimental data support the model, a key benefit is that we can draw on many already-existing sciences, including physics, chemistry, biology, and ecology.

This book aims to be practical, providing measurement methods and equations that can be coded. My long term aim is to grow justice, equity, creativity, and dignity for people all around the world. The book is far from perfect, and contains many errors, inconsistencies, and gaps. I am sorry for these shortcomings, but I have balanced them with a desire to get the book quickly to my students in the CH E 497B: Physics of Community course, and to my colleagues. And so I am also sorry for the delay! If the ideas have merit, I am confident that we collectively will make repairs. I invite critique, collaboration, and experiments.

For my students, I hope the notes convey the essence of the ideas, so that you can change the world with them, since you are inheriting a world that requires better decision making processes.

Enjoy!

Darrell Velegol
State College, Pennsylvania
2015 October 31

Physics of Community Course Notes for Fall 2015

Quantifying collective decisions in a world of power, dialogue, conflict, selfishness, corruption, altruism, and creativity.

Darrell Velegol

Wild Scholars Media

1. Overview. PC and quantitative collective decisions

... just as game theory without broader social theory is merely technical bravado, so social theory without game theory is a handicapped enterprise. – Herbert Gintis[1]

Collective decisions **lack a clear "best".**

Communities consist of two or more people, who interact together in some space and time. Frequently the space and time are such that resources are scarce, and decisions must be made about how to allocate scarce resources with alternative uses. Making these strategic decisions is the subject of economics or political science. Mathematical techniques can be used to make allocations to give an optimum or best solution among the participants.

In introductory calculus courses, we learn how to optimize (i.e., maximize or minimize) a simple function. Here is an example:

Example 1-1. Maximizing a function of one variable.

Find the maximum of the function $y = 6 + 4x - 2x^2$.

SOLUTION. By inspection of the figure below, we see that the maximum occurs at x = 1, and at a value of y = 8. To find the value analytically, we take dy/dx = 0 = 4 − 4x, and thus find the same position for the maximum that inspection gave. Substituting x = 1 into the function gives y = 8, as expected from inspection. We can check to find that the second derivative gives d^2y/dx^2 = -4, with the negative value consistent with a maximum.

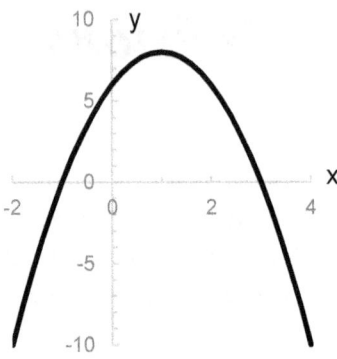

More complicated optimizations occur when we seek to optimize an objective function, subject to constraints. Example 1-2 gives a simple example, which we can solve using classic linear programming methods.[2,3]

Example 1-2. Maximizing an objective function, in the presence of constraints. This example gives a linear programming problem.

We have two fractions x and y that we can control. We want to minimize a function (the objective function) z = 70x + 50y. There are a few constraints on this minimization. Both x and y must be non-negative, and they must add to 1 (i.e., total of fractional choices = 1). In addition, we require that 0.20x + 0.30y ≤ 0.28. We can rewrite this problem in a standard form as min $z = 70x + 50y$

s.t. $0.20x + 0.30y \le 0.28$

$x + y = 1$

$-x \le 0, \quad -y \le 0$

SOLUTION. This is a classic linear programming problem, readily solved even in spreadsheet software like Excel. Coding this into Excel gives x = 0.20, y = 0.80, and z = 54. All the constraints are satisfied, and we see that for any change in x or y, the value of z becomes larger. We can also draw out the solution graphically, as shown below. The four constraints are plotted as lines. They collectively give a "feasible region". The theory of linear programming tells us that the minimum for z will occur at one of the boundary points of the region, highlighted by the small circular dots. Upon trying each of the dots, we find that the unfilled dot gives the minimum value of z. The theory of linear programming can be used to solve for strategies for constant sum strategic games, which most commonly means zero-sum games.[2]

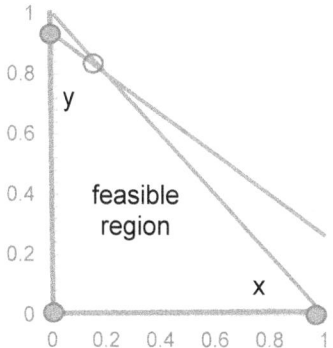

In both of these examples, a clear definition of "the best" exists. But what happens if two or more people have different definitions of "the best"? Here starts the problem examined in this book: How do we make collective choices when our individual visions of "the best" are different, either in kind or in degree?

Collective decisions give rise to politics and power.

Here I mention a few of the many arenas in which collective decisions occur. As a Chemical Engineering faculty, I see a lot of amazing research in the biomolecular, nanomaterials, and computational sciences, which can be applied to critical problems of our day. However, I see collective decisions as being a primary bottleneck to the implementation of sound technical solutions. The challenge is this: Oftentimes when even a small fraction – or even one person – of a group will be *disadvantaged*, that person will fight to hold their position of relative advantage. As a result, collective decisions involve competition, conflict, creativity ... or even corruption! If handled properly, the conflict can be beneficial to all parties. In this book we will examine a variety of collective decisions, from the Perspectives of the various Players. Table 1-1 gives a sample of strategic decisions, and Players who might be involved in making the decisions.

Table 1-1. Sample of strategic decisions, and Players who might be involved. Some decisions are fleeting, some are quite significant. Many of the decisions in this table are made in joint pairs.

arena	collective decision	possible Players
career	Who gets promoted? Who has to work weekends? Who gets to work weekends? How much to brown nose? Who gets the awards?	boss, employee, Dilbert
corporation	Which product launches? How to allocate investments? How to clean up this oil spill? Who to blame for this explosion?	CEO, VP of R&D or finance or marketing
family	Who does the laundry?	parents, kids

	Who will cook dinner? Who will clean up after dinner? Who will shop for groceries? Who will walk the dog? Who will inherit the money?	
marriage	How will we spend the finances? How many kids will we have? How often will we have sex?	partners
town	How to deal with neighbors' dog? How to reduce sexism and racism?	Mayor, community organizer
school	Should I give an easy exam? Should I study hard? Matching students-teachers? Should I curve this course? College tuition rates? Do I obey the bulley? Whom should I date?	principal, teacher, students, school board members, board of trustees members
sports	Who makes the team? Allow NCAA pay? Soccer championship location?	coach, players, NCAA president, athletic director, FIFA president
society	CO_2 taxes for climate change? Immigration laws? Tax rate? Austerity versus spending? NSF funds distribution?	Speaker of the House, NSF Director, Senate majority leader
medicine	Who gets the kidney? Will the insurance pay for this?	doctors, patients, donors
transportation	Which infrastructure to repair? Which routes to secure? What safety measures to promote?	state governor, state majority leader, AAA president, CEO of State Farm
climate	Support renewable energy? Use more natural gas or less?	state governor, head of EPA

Strategic decisions are quantified with game theory.

Many important books have been written about strategy, power, influence, cooperation, competition, and similar topics. Some of my favorites are by Pfeffer,[4] Gaventa,[5] Boulding,[6] Dahl,[7] Alinsky,[8] Horton,[9,10] and Freire[11]. These works range from qualitative to quantitative. The qualitative books, such as those by Gaventa and Boulding, have a useful "3-dimensions" or "3-faces" of power taxonomy. These important qualitative works do not enable the quantitative design of decisions that this book seeks to establish, although they are enormously important books that will provide guidance for this book.

In 1943 John von Neumann and Oskar Morgenstern showed a simple way to represent competitive decisions for "rational players".[12] Their book introduced us to "game theory", which portrays information about the contest in a tabular or graphical form.[1,13,14,15,16] Examples 1-3 and 1-4 gives two classic examples from game theory, for the Prisoners Dilemma game and the Battle of the Sexes game.

The "solutions" to a game are given whenever we find a "Nash equilibrium". John Nash, Nobel Prize winner in Economics in 1994, showed that there will always be at least one equilibrium in a finite game, although there can be several, even for a 2×2 game, as we will see in Example 1-4. In a Nash equilibrium both Players know the others' strategy – i.e., the fraction of time that Player A chooses a1 (f_{a1}) versus a2 (f_{a2}), and Player B chooses b1 (f_{b1}) versus b2 (f_{b2}). Equilibrium occurs when, even knowing the strategy of the other Player, neither Player has an incentive to change his or her own strategy unilaterally. This can sound confusing, and so a couple examples will clarify this.

Example 1-3. The Prisoners Dilemma (PD) game.

Stephanie (Player A) and Darrell (B) stand outside the jewelry store, ready to burgle it. Just then, an alert police officer arrests them, separates them, and takes them to the police station, where he places them in separate cells. The officer urges them to tell on each other, and conveys what will be the punishment if they remain "quiet", or "tell" (our two choices). Since each Player has two choices, there are four possible outcomes. For example, if Stephanie (A, bold italics) tells (choice a2) and Darrell (B) remains quiet (choice b1), then Stephanie gets zero sentence, while Darrell gets a sentence of 7 months in jail. If Darrell tells while Stephanie stays quiet, then Stephanie gets 7 months in jail, and Darrell gets 0. The officer really wants both to tell, and so if one tells and the other stays quiet, he induces both to tell by giving 5 months to each. And in the end, if both remain quiet, the officer will give them each 2 weeks on a trumped-up loitering charge! What should each Player do?

SOLUTION. First, we seek to represent the decision process, using the classic "normal form" from game theory. We label the Possibilities for each Player as a1, a2, b1, and b2. In the cells we list the penalty or Pain that each will receive. For now, these are listed simply in terms of the number of months in prison. The values for Player A are listed in **bold Italic**, while those for Player B are listed as ordinary font. Thus, if A chooses a1 and B chooses b2, then A receives a 7.0 month prison term, while B receives 0.0.

	b1 = quiet	b2 = tell
a1 = quiet	**0.5**, 0.5	**7.0**, 0.0
a2 = tell	**0.0**, 7.0	**5.0**, 5.0

In this book we will call this normal form game the *Pain matrix* or the game marix.

Let's say for the moment that B chooses b1 (quiet). We see that A does better if she tells (a2), since then she receives a lighter prison sentence (0.0 months, instead of 0.5). What if B tells (b2)? We see that still, A does better to tell (a2) than to remain quiet (5.0 months compared with 7.0 months). That is, in either case, A does better to tell (a2).   A similar analysis for B shows that he likewise should always tell (b2).

The strategy for this game is a *pure strategy*,[1] which we will represent here as $\{f_{a1}, f_{b1}\} = \{0.0, 0.0\}$.   We could also have represented the strategy as $\{f_{a1}, f_{a2}, f_{b1}, f_{b2}\} = \{0.0, 1.0, 0.0, 1.0\}$ or $\{f_{a2}, f_{b1}\} = \{1.0, 1.0\}$.   The "dilemma" here is that since both Players tell, they both receive a worse punishment than if both would remain quiet! That is, the best individual choice gives a worse result, even for the individual, than the best collective choice! The Prisoners Dilemma game has been compared to results for climate change, fisheries,[17] and other tragedy of the commons[18] problems.

We will add one final piece to this classic game, which are the *expected pain matrices*.   These are calculated using mathematical expectation theory.   For example, if there is a 30% probability of receiving 5 months and a 70% probability of receiving 7 months, then the mathematical expectation is for a sentence of (0.30)(5 months) + (0.70)(7 months) = 6.4 months.   If we look at both Players, the probability of attaining the cell for (a1,b2) is $f_{a1}f_{b2} = f_{a1}(1-f_{b1})$, since f_{a1} + f_{a2} = 1.   Therefore, the mathematical expectation result (E) for either Player, given the probability $\{f_{a1}, f_{b1}\}$, is given as

(1-1)

$$
\begin{aligned}
E_A &= f_{a1}f_{b1}p_{A11} + f_{a1}\left(1 - f_{b1}\right)p_{A12} \\
&\quad + \left(1 - f_{a1}\right)f_{b1}p_{A21} + \left(1 - f_{a1}\right)\left(1 - f_{b1}\right)p_{A22} \\
E_B &= f_{a1}f_{b1}p_{B11} + f_{a1}\left(1 - f_{b1}\right)p_{B12} \\
&\quad + \left(1 - f_{a1}\right)f_{b1}p_{B21} + \left(1 - f_{a1}\right)\left(1 - f_{b1}\right)p_{B22}
\end{aligned}
$$

where the $p_{\alpha ij}$ values are given in the Pain matrix for Player α, for the cell (a_i, b_j).   The Table below shows these expectation values for Players A and B, with probabilities f_{a1} on the vertical, and f_{b1} on the horizontal.   For example, if Player A has a 60% probability of selecting a1, and Player B has an 80% probability of selecting b1, then the expected Pain for Player B is +2.88.

---

[1] When we get to Gibbsian games in later chapters, we will see that "pure strategies" do not occur. Some strategies might happen very infrequently, but they still have a finite probability of occurrence.

A: fa1\fb1	0.00	0.20	0.40	0.60	0.80	1.00
0.00	5.00	4.00	3.00	2.00	1.00	0.00
0.20	5.40	4.34	3.28	2.22	1.16	0.10
0.40	5.80	4.68	3.56	2.44	1.32	0.20
0.60	6.20	5.02	3.84	2.66	1.48	0.30
0.80	6.60	5.36	4.12	2.88	1.64	0.40
1.00	7.00	5.70	4.40	3.10	1.80	0.50

B: fa1\fb1	0.00	0.20	0.40	0.60	0.80	1.00
0.00	5.00	5.40	5.80	6.20	6.60	7.00
0.20	4.00	4.34	4.68	5.02	5.36	5.70
0.40	3.00	3.28	3.56	3.84	4.12	4.40
0.60	2.00	2.22	2.44	2.66	2.88	3.10
0.80	1.00	1.16	1.32	1.48	1.64	1.80
1.00	0.00	0.10	0.20	0.30	0.40	0.50

We see that for A (upper table), no matter what choice B makes, A minimizes Pain by playing the strategy $f_{a1} = 0$ – i.e., always play a2. Similarly for B (lower table), no matter what choice A makes, B minimizes Pain by playing the strategy $f_{b1} = 0$ – i.e., always play b2.

Example 1-4. The Battle of the Sexes (BoS) game.

Stephanie (Player A) and Darrell (B) plan to go out for a date. Stephanie wants to go on a hike, while Darrell wants to see a movie. But above all, they want to be together. The game matrix for this situation is given below; I have not made it symmetric, just to show that there is no need for it to be so. For reasons that I will clarify later, we will keep "pain" as a positive number, and "pleasure" as a negative number.[19] We want to find the Nash equilibria for this situation.

	b1 = hike	b2 = movie
a1 = hike	*-3.5*, -2.3	*-0.2*, -0.4
a2 = movie	*+0.3*, +0.5	*-2.6*, -3.3

SOLUTION. Unlike in the Prisoners Dilemma example, there is no dominant strategy solution for this situation. If B chooses b1, then A would choose a1, since it has a lower (i.e., more negative, and therefore more pleasurable) pain. But if B chooses b2, then A would choose a2. In this situation, one actually finds 3 Nash equilibria!

One of the equilibria occurs when $\{f_{a1} = 1.0,\ f_{b1} = 1.0\}$. If the Players choose this strategy, there is no reason for either of them to change their strategy individually.   Likewise, another Nash equilibrium is $\{f_{a1} = 0.0,\ f_{b1} = 0.0\}$. Again, if the Players choose this strategy, neither would change unilaterally. These are both pure strategy solutions, since both Players choose to hike, or both Players choose to go to the movie.

There is a third strategy solution for this problem. This one is called a "mixed strategy", because each Player will choose to hike with some non-zero and non-unity probability. That is, each will choose to hike with some finite probability, and each will choose to attend the movie with some finite probability. We want to find the point where the Players are indifferent, such that even though they know the strategy of the other – even though they don't know the choice on any given play of the game – neither player wants to change.

For the mixed strategy solution, when one exists, we require that the Pain that A feels does not depend upon f_{a1}. Likewise, we require that the Pain that B feels does not depend upon f_{b1}. Mathematically we express these conditions as

(1-2)
$$\frac{\partial E_A}{\partial f_{a1}} = 0, \quad \frac{\partial E_B}{\partial f_{b1}} = 0$$

But now we can insert Eq 1-1 into Eq 1-2 to obtain

(1-3)
$$E_A = f_{a1} f_{b1} p_{A11} + f_{a1} (1 - f_{b1}) p_{A12}$$
$$+ (1 - f_{a1}) f_{b1} p_{A21} + (1 - f_{a1})(1 - f_{b1}) p_{A22}$$
$$E_B = f_{a1} f_{b1} p_{B11} + f_{a1} (1 - f_{b1}) p_{B12}$$
$$+ (1 - f_{a1}) f_{b1} p_{B21} + (1 - f_{a1})(1 - f_{b1}) p_{B22}$$

Substituting Eqs 1-) into Eq 1-2, we attain the following equations:

$$f_{a1} = \frac{P_{B22} - P_{B21}}{P_{B11} - P_{B12} - P_{B21} + P_{B22}}$$

(1-4)

$$f_{b1} = \frac{P_{A22} - P_{A12}}{P_{A11} - P_{A12} - P_{A21} + P_{A22}}$$

These equations apply fairly broadly for "mixed strategy" solutions.[20] That is, when the solutions from Eq 1-4 are $0 \leq f_{ai} \leq 1$ for Player α, Possibility i, then Eq 1-4 gives the solution. If the f values are outside this range, then a dominant or dominated strategy has entered the picture.

NOTE: In Eq 1-4, the result for Player A depends solely on the Pains for Player B! This is a counter-intuitive result! It says that each Player makes a decision based on what the other Player will accept.

Using Eq 1-4 for this BoS game, we find that a third Nash equilibrium occurs at $f_{a1} = 0.667$ and $f_{b1} = 0.387$ (to 3 decimals). Entering the exact values into the expected pain matrix shows the mixed strategy solution in the expected pain matrix below. If in fact A plays with $f_{A1} = 0.667$ and B plays with $f_{B1} = 0.387$, then the pain to A is -1.4777, and to B is -1.367. That is, both Players are in negative pain, or pleasurable, outcome. Now let's say that Player B tries to have an even lower pain (more pleasurable) situation. Why do we call this an "equilibrium"? If either A or B were to Play any other value, neither would personally gain or lose, and so neither has an incentive to change.

A: fa1\fb1	0.000	0.200	0.387	0.600	0.800	1.000
0.000	-2.600	-2.020	-1.477	-0.860	-0.280	0.300
0.200	-2.120	-1.788	-1.477	-1.124	-0.792	-0.460
0.400	-1.640	-1.556	-1.477	-1.388	-1.304	-1.220
0.667	-1.000	-1.247	-1.477	-1.740	-1.987	-2.233
0.800	-0.680	-1.092	-1.477	-1.916	-2.328	-2.740
1.000	-0.200	-0.860	-1.477	-2.180	-2.840	-3.500

B: fa1\fb1	0.000	0.200	0.387	0.600	0.800	1.000
0.000	-3.300	-2.540	-1.829	-1.020	-0.260	0.500
0.200	-2.720	-2.188	-1.690	-1.124	-0.592	-0.060
0.400	-2.140	-1.836	-1.552	-1.228	-0.924	-0.620
0.667	-1.367	-1.367	-1.367	-1.367	-1.367	-1.367
0.800	-0.980	-1.132	-1.274	-1.436	-1.588	-1.740
1.000	-0.400	-0.780	-1.135	-1.540	-1.920	-2.300

As with the PD game, having a Nash equilibrium does not mean that there is a Pareto optimum. Indeed, both Players would be better off, if they chose either of the "corner solutions" listed earlier.

Classical game theory has critical shortcomings.

The two examples in game theory might suggest that we now have the key tool to examine social science problems. Of course, many extensions are needed to the simple game theory shown, but many modifications have already been worked out.[15,16] These modifications include repeated games, coalitions (collections of players that take sides), sequential games, and many more. Fundamentally, we might think that classical game theory can solve all our social problems.

And yet, there are numerous shortcomings to game theory. A significant shortcoming is described by Herbert Gintis,[21] who states in his book *The Bounds of Reason*:

> *The most fundamental failure of game theory is its lack of a theory of when and how rational agents share mental constructs .... Humans have a social epistemology,*

meaning that we have reasoning processes that afford us forms of knowledge and understanding, especially the understanding and sharing of the content of other minds, that are unavailable to merely "rational" creatures .... The bounds of reason are thus not the irrational, but the social.

This failure of game theory essentially focuses on the inputs to the game, and seems related to Herbert Simon's concept of "bounded rationality". And indeed, this is a significant shortcoming. The shortcomings I list below are for how the games are solved and interpreted. I list some of the most prominent, from the perspective of a Chemical Engineer.

1. *set of solutions.* Even simple games have multiple Nash equilibria – a set of strategy solutions – and there is no clear way to determine how often one strategy will be played, compared with another. This does not happen with chemical equilibrium.

2. *non-continuous solutions.* Part of this is related to dominant and dominated solutions. The concept is easy to see with the example games given below. Classical game theory gives the same solution to the two Prisoner Dilemma games below. If the Players were to deviate from {tell, tell}, they would be deemed "irrational". But do we really pay much attention, for example, if a half gallon of milk costs $1.96 or $1.98?

	b1 = quiet	b2 = tell
a1 = quiet	*0.5*, 0.5	*7.0*, 0.0
a2 = tell	*0.0*, 7.0	*5.0*, 5.0

	b1 = quiet	b2 = tell
a1 = quiet	*4.99*, 4.99	*5.01*, 4.98
a2 = tell	*4.98*, 5.01	*5.00*, 5.00

3. *rationality*. Beyond Herbert Simon's concept of "bounded rationality", in which a person cannot collect enough data or compute the outcome, we usually visualize "rational" only in the sense that it helps oneself! Thus, there is no real altruism, revenge, win-win, or fair behavior, except in light of self-interest. Here I much prefer Gintis' definition of rational: "A rational actor is an individual with consistent preferences."[22] Gintis then describes his beliefs-preferences-constraints (BPC) model of rationality.

4. *spurious tipping points*. In Example 1-4, we found an interior solution $\{f_{a1} = 0.667$ and $f_{b1} = 0.387\}$. If B decides to play at $f_{b1} = 0.386$ – just a tiny difference – then A would immediately change to $f_{A1} = 0$. This seems like an excessively dramatic "tipping point" change for A, given the almost infinitesimal change from B. Such tipping points are common in classical game theory.

5. *utility theory*. There are numerous utility theories, from expected value (utility proportional to the amount of the good)[,12] expected utility (percent change in utility proportional to the percent change in the good, from Bernoulli), Cobb-Douglas (a production function using a power law), to prospect theory (which accounts for diminishing returns and reference points). Estimating utility from any of these theories is challenging.

There are thus many features that we might prefer to have in game theory:

1. *true equilibrium*. Chemical equilibria tend to one stable state, even though there can be metastable regions. The critical factor missing in classical game theory is "entropy". Entropy moves

us from "mathematical expectation" to "Gibbsian expectation". We would thus like to include both energetic and entropic aspects, allowing solutions to move continuously, rather than in spurious tipping point jumps. To do this, we will treat ideas and alternatives as a type of molecule, which we will call a "knowlecule".

2. *satisficing.* There is no version of "kT" (per molecule) or RT (per mole) in game theory at present, which is really needed for the Herbert Simon concept of "satisficing". We would prefer that pains be measured by kT or RT.

3. *creativity, crime, and no-play.* Classical game theory assumes that one of the alternatives for each Player must be played, and furthermore that no additional ones are possible. Classical game theory therefore does not usually allow for creative responses. Nor does game theory introduce the option of various corruption or crime in each decision, nor does it usually offer the possibility of "no-play", when none of the options are acceptable.

PC addresses the shortcomings in a consistent framework.

This book proposes a quantitative and consistent framework for collective decisions. The hypothesis is that humans process ideas and alternatives, and then make decisions, in a manner similar to how a chemical plant processes reactant molecules to form product molecules. In a very interesting paper by Veloz et al., parts of this concept have been called the "socio-chemical metaphor".[23] Numerous authors have examined economic principles in light of physical principles, and as discussed by Mirowski, these attempts have mostly ended up as "more heat than light".[24] The quote in the preface by Nobel laureate Paul Samuelson expresses his opinion.

This book extends the socio-chemical concept from reaction kinetics, to reaction equilibrium, and adds the processing steps of unit operations ("unit ops"). In the field of Chemical Engineering, designs are examined at three levels: the systems level (large, with unit operations), the molecular level (small, with atoms and molecules and chemistry), and the continuum level (in between, with continuum equations). Table 1-2 describes these levels, and relates them to other disciplines.

Table 1-2. Three levels of Chemical Engineering analysis, compared with collective decisions. Notice the many P's.

level	Chemical Engineering description
process (Ch 4) Identify a <u>Problem</u> to solve, the <u>Players</u> (i.e., their "unit ops"), and <u>Process connections</u> and relationships into a Process Flow Diagram (PFD).	The key pieces of equipment, such as reactors, separators, energy exchangers, and pumps are drawn as a block flow diagram (BFD), or more precisely, as a process flow diagram (PFD). This stage is often done heuristically, and gets the large "unit operations" in place and connected. The largest plants are more than 1000 acres in size. PC draws upon network theory for parts of the analysis.
chemistry (Ch 2) Identify <u>Possibilities</u> (choices) and their <u>Prior</u> (starting) amounts), <u>Products</u> (decision outcomes), <u>Pains</u> for each outcome, and <u>Perspectives</u> (e.g., selfish, altruistic).	At the smallest scale, particular chemical reaction "stoichiometry" is written. For example, the reaction $2NaCl + H_2O \rightarrow Cl_2 + H_2 + 2\ NaOH$ is for the production of chlorine. Although quantum mechanics determines feasible chemical reactions, usually reactions are considered symbolically by their letters. There are just over 100 elements like chlorine (Cl), sodium (Na), oxygen (O), and hydrogen (H). PC draws upon and expands game theory for analysis, treating choices as "knowlecules".
continuum (Ch 3) Calculate the <u>Probabilities</u> for the outcomes from the unit ops, and the entire Process.	In between PFDs and molecular reactions, analysis is done by continuum level equations. These dictate the first and second laws of thermodynamics, Newton's law, conservation of mass, as well as rate equations for heat transfer, mass transfer, chemical reactions,

	and other phenomena. PC also draws upon information and utility theories for analysis.

In this book we will quantitatively relate strategic decisions metaphorically to these three levels. The analysis will reveal many assumptions in economic and political theory, which in any chemical engineering design must be addressed before a chemical plant can be commissioned. But there are important pieces, which even though some have been touched on in the literature, they have not been collected into a coherent whole, analogous to that needed to design a chemical plant. And so for the process level, we will examine not so much the usual network theory problems of "shortest path" or "critical path analysis", but I will examine mixing concentrations of information, contamination, purification, and flowrates. In addition, I "separate people" or agents into their "unit op parts", such as creator, evaluator, encourager. For the chemistry level, I introduce classical game theory and Nash equilibria, but I focus toward "Gibbsian games" and "entropic choices". These methods lead us to envision choices as "molecular entities" – I use the word "knowlecules" – which can be represented symbolically. For the continuum level, I discuss information theory and information transmission, but perhaps even more so focus on information creativity, transformation, and memory, and include an information enthalpy.

As with all physical systems, the logic in this book is often subtle. The following quote comes from Dill & Bromberg:

The challenge in understanding these behaviors [chemical phenomena] is that the properties that can be measured and controlled, such as density, temperature, pressure, heat capacity, molecular radius, or equilibrium constants, do not predict the tendencies and equilibria of systems in a

> *simple and direct way. To predict equilibria, we must step*
> *into a different world, where we use the language of*
> *energy, entropy, enthalpy, and free energy.*[25]

The physics is not Newtonian or quantum physics, but Gibbsian physics, which is by nature chemical and statistical.

In chemical physics, the methods are abstract, and yet rigorous and predictive. One might say, "The model is too complex, and besides, you can fit anything with so many parameters." However, in reality the results in this book are what are now used to predict systems, and indeed a simplified version of that.

In this work I treat decisions as "decision reactions between choices", where the choices can come from any number of Players. These decisions and pieces of information in turn flow to other Players, who then use them for their own decision making process. The hypothesized view of strategic decisions has similarities to ecology or biology, in which every source processes information and acts as a predator or as food for the next form of life!

Let's look at an example in which we can see how the three levels of analysis enter. This example describes a choice that is made every semester by students.

Example 1-5. Purchasing a college textbook. This simplified story is one that we will use later in the book again.

Jill is a college junior enrolled in a course that "requires" a certain book. The campus bookstore manager is deciding the cost of the book, setting a market for the book. The manager knows the costs: The bookstore has paid $15 to buy the book wholesale, and if the book will take its share of fixed store costs, it would cost $20, and with operating expenses would cost $24.   The bookstore would prefer to make a small profit on the book, to cover other expenses, and to bring store upgrades. But the manager also knows the realities: If the cost is too high, students might well just go out for pizza with their friends, instead of buying the book! Some simple research estimated that students drop by the bookstore with $40 in their pockets, and it costs roughly $14 for pizza and a movie. What should the manager charge, and will the student buy?

SOLUTION. The problem statement gives one version of the story. We might want more background on Lauren, about her academic commitment for instance. We might want to know some historical data about pricing, and the fraction of students who buy the books. We don't have it yet.

At the Process Level, we know that the bookstore will make their decision first, by choosing a price. That information will be based on the internal decisions of the manager, with her current knowledge of prices and history. The price information will flow forward to Jill, who as a junior has experience with purchasing books, knowing typical prices, and perhaps has information about how much the professor will use the book. We also know that the bookstore provides a market for the book, enabling its sale.

At the Chemistry Level, we start to examine the choices. The manager could in principle charge any price, including giving the book along with giving money (i.e., a negative price), up to $1000s. She could choose $25.02, $25.03, or $25.04. In practice, the manager might choose $20, $25, or $30, based on her knowledge as given in the story. Jill also has choices. She could choose to purchase the book at the given price, or forego the book and instead go for pizza and a movie. She might also have a creative solution, such as checking on Amazon for the book. Or she might have a corrupt solution, trying to steal the book. Let's simplify the matrix to the following, where A = Jill and B = manager. We will choose the Pains arbitrarily at this point, although later we will use a method to estimate the Pains. The manager does not want to be stuck with the book, and has some pain if she still has it after Jill makes a decision.

	b1 = $20	b2 = $25	b3 = $30
a1 = book	**-1.5**, +2.0	**-1.3**, -1.0	**-1.1**, -2.0
a2 = movie	**-1.0**, +1.0	**-1.0**, +1.0	**-1.0**, +1.0
a3 = both	**-2.3**, +2.0	**-2.0**, -1.0	**0.0**, +1.0

We have thus written the Players, their Possibilities, and their Pains, and the Problem is to calculate the Probabilities for their choices (i.e., their strategies). We can assume that their Perspectives are purely self-interested at this point, each wanting to maximize their own position, since the intermediary is the market. Using results from classical game theory – or later using Physics of Community (PC) principles – we can calculate the Probabilities for each of the 9 possible decision outcomes. Having this ability then allow us to examine various design scenarios for the pricing, and other creative opportunities.

Some of the new concepts I will explore in this book include a more full application of the "socio-chemical metaphor" mentioned before, the use of a "pain potential" to match the chemical potential of chemical physics, "Gibbsian choices" which include both the enthalpy and entropy of decisions (and so includes entropic choices), as well as the P's mentioned above in Table 1-2.

Symbols

A, B, C ... = Players in the games
$a1, a2, a3$ ... $b1$ ... $c4$ ... = Possibilities for choices for Players
f_{a1}, f_{a2} ... f_{b1} ... $f_{\alpha i}$ = fraction of time that Player α chooses their ith Possibility

Exercises

1. *optimizing a function.* Find the extrema of the function $y = 2x^3 - 5x^2 + 3x - 14$.

2. *linear programming.* Use Excel to solve the linear programing problem given in Example 1-2.

3. *mixed strategy derivation.* Expand the derivation in the book, to derive Equation 1-4.

4. *chicken game*. Solve for the strategies of the "Chicken game", in which two drivers play the game of chicken. Also give the probable pain matrix, for each Player choosing to "swerve" 0%, 25%, 50%, 75%, and 100% of the time. The pains are given in the table below. See https://en.wikipedia.org/wiki/Chicken_%28game%29 for details.

	b1 = swerve	b2 = straight
a1 = swerve	**0.00**, 0.00	**+1.00**, -1.00
a2 = straight	**-1.00**, +1.00	**+10.00**, +10.00

5. *rock-paper-scissors (RPS)*. Using a similar method as for the BoS game, derive a result for the strategies the Players should use, for the 3×3 RPS game. The key part is setting up the pain (E_A) as a function of f_{a1}, f_{a2} and f_{a3} for instance, and then taking

$$\frac{\partial E_A}{\partial f_{a1}} = 0, \quad \frac{\partial E_A}{\partial f_{a2}} = 0$$

As before, if there is no change with a_1 or a_2, there will be no change with a_3. Then you do the same with Player B. For the pains, set a "win" (e.g., paper covering rock) as -1, a loss at +1 pain, and a null at 0. Then calculate the strategy of either Player, which is {fraction of time playing rock, fraction for paper, fraction for scissors}. You should find the following mixed strategy solutions:

$$f_{a1} = \frac{\left(p_{B33} - p_{B31}\right)\left(p_{B22} - p_{B23} - p_{B32} + p_{B33}\right) - \left(p_{B33} - p_{B32}\right)\left(p_{B21} - p_{B23} - p_{B31} + p_{B33}\right)}{D_A}$$

$$f_{a2} = \frac{\left(p_{B33} - p_{B32}\right)\left(p_{B11} - p_{B13} - p_{B31} + p_{B33}\right) - \left(p_{B33} - p_{B31}\right)\left(p_{B12} - p_{B13} - p_{B32} + p_{B33}\right)}{D_A}$$

$$f_{a3} = 1 - f_{a1} - f_{a2}$$

$$D_A = \left(p_{B11} - p_{B13} - p_{B31} + p_{B33}\right)\left(p_{B22} - p_{B23} - p_{B32} + p_{B33}\right)$$
$$- \left(p_{B21} - p_{B23} - p_{B31} + p_{B33}\right)\left(p_{B12} - p_{B13} - p_{B32} + p_{B33}\right)$$

$$f_{b1} = \frac{\left(p_{A33} - p_{A13}\right)\left(p_{A22} - p_{A23} - p_{A32} + p_{A33}\right) - \left(p_{A33} - p_{A23}\right)\left(p_{A12} - p_{A13} - p_{A32} + p_{A33}\right)}{D_B}$$

$$f_{b2} = \frac{\left(p_{A33} - p_{A23}\right)\left(p_{A11} - p_{A13} - p_{A31} + p_{A33}\right) - \left(p_{A33} - p_{A13}\right)\left(p_{A21} - p_{A23} - p_{A31} + p_{A33}\right)}{D_B}$$

$$f_{b3} = 1 - f_{b1} - f_{b2}$$

$$D_B = \left(p_{A11} - p_{A13} - p_{A31} + p_{A33}\right)\left(p_{A22} - p_{A23} - p_{A32} + p_{A33}\right)$$
$$- \left(p_{A12} - p_{A13} - p_{A32} + p_{A33}\right)\left(p_{A21} - p_{A23} - p_{A31} + p_{A33}\right)$$

6. *biased rock-paper-scissors (RPS).* Using the results from the previous problem, calculate the result for the usual RPS game, and for the "peaceful RPS game" below. Note that the usual RPS game should give 1/3 for f_{a1}, f_{a2}, f_{b1}, and f_{b2}.

	b1 = R	b2 = P	b3 = S
a1 = R	**0**, 0	**0**, -4	**-1**, +1
a2 = P	**-4**, 0	**0**, 0	**0**, -1
a3 = S	**+1**, -1	**-1**, 0	**0**, 0

Note that since "paper" is the "most peaceful" Possibility, its value is the highest (0.5). However, since the Players will use more paper, the amount of scissors will also rise (0.4). This leaves rock as the least popular Possibility (0.1).

7. *textbook purchase*. Use the 3×3 solution given above to solve Example 1-5 about the textbook purchase. Note that the solution applies only if both f values are 0 to 1. Check each block to find two pure strategy Nash equilibria.

	b1 = $20	b2 = $25	b3 = $30
a1 = book	**-1.5**, +2.0	**-1.3**, -1.0	**-1.1**, -2.0
a2 = movie	**-1.0**, +1.0	**-1.0**, +1.0	**-1.0**, +1.0
a3 = both	**-2.3**, +2.0	**-2.0**, -1.0	**0.0**, +1.0

8. *voting (challenge problem)*. Using a similar method, set up the derivation for solving a voting game with 3 Players and 2 choices each. This is a more challenging problem than 2 Players with 3 choices each, since this problem becomes nonlinear. If you are able to solve it, do you see a coalition form? Is there a more stable state with a coalition?

References

Fudenberg, Drew; Tirole, Jean. *Game Theory* (1991). This is the "grown up version" of game theory, written by masters of the craft. Tirole won the Nobel Prize in Economics in 2014.

Gintis, Herbert. *The Bounds of Reason: Game Theory and the Unification of the Behavioral Sciences*, revised Edition (2009). This book is among my favorites, because it addresses not only the technical calculations of game theory, but also some philosophical underpinnings.

Leyton-Brown, Kevin; Shoham, Yoav. *Essentials of Game Theory: A Concise Multidisciplinary Introduction* (2008).   This is another nice introduction to game theory, written at a slightly more sophisticated level than Spaniel's book above.   In addition, Leyton-Brown and Shoham teamed up with Matt Jackson to offer a nice MOOC on Game Theory.

Osborne, Martin J.; Rubinstein, Ariel. *A Course in Game Theory* (1994). This book is written at an intermediate level.

Spaniel, William. Game Theory 101: The Basics (2011). This is a simple, short, nice introduction to game theory. I always tell my own students to start with this book. Spaniel has also written a more full text *Game Theory 101*, as well as an interesting book on *War*.

von Neumann, John; Morgenstern, Oskar. *Theory of Games and Economic Behavior*. Princeton (1944). The seminal book on game theory.

2. Chemistry. Decision reactions and stoichiometry.

I formed the opinion that, even though much of the recent progress in structural chemistry has been due to quantum mechanics, it should be possible to describe the new developments in a thorough-going and satisfactory manner without the use of advanced mathematics. – Linus Pauling[26]

Decision reactions have a stoichiometry.

In the field of chemistry we can write the combustion of octane, an important component of gasoline, as follows:

$$C_8H_{18} + 12.5O_2 = 8CO_2 + 9H_2O$$

This chemical equation states that if I add 1.0 part octane with 12.5 parts oxygen by number, then these reactants will form products consisting of 8.0 parts carbon dioxide and 9.0 parts water. The species and coefficients represent the stoichiometry of the reaction, a fundamental concept in chemistry. There are several parts of this relatively simple chemical equation that I find astonishing:

- symbols. We symbolize "octane" by C_8H_{12} and water by H_2O; however, these chemical species contain many electrons, protons, and neutrons, distributed in space in complex orbitals, all described by quantum mechanics. The description of even the simplest atom, hydrogen (H), is quite involved, requiring extensive study of quantum mechanics to understand. A molecule like octane is much more complicated. And yet, writing a few simple symbols to describe extraordinarily complex entities is useful and works quite well.

- proportions. The proportions of molecules required in the reaction are predictable and consistent. That is, they have a set stoichiometry. These proportions are one manifestation of the law of conservation of mass, which says that matter is neither created nor destroyed. The conservation of mass law holds extraordinarily well in all situations, expect during nuclear reactions. Of course, some stoichiometries are hard to determine, such as in pyrolysis or combustion.

- elements. There are 98 naturally-occurring elements, such as hydrogen (H), oxygen (O), gold (Au), and others. All of the complexity of soil, rocks, trees, humans, forests, oceans … starts with about 100 elements.

In this chapter I use simple symbols to represent choices available to a decision maker. If we back up, we see that normal form games provide a useful, compact representation of a game. In this Chapter, however, I will introduce a different representation, which has been called the "socio-chemical metaphor" or framework. We will write normal form games as sets of chemical reactions, which I call decision reactions, which will enable us to use the powerful apparatus of chemical reaction network theory (CRNT[27]) and other highly-developed tools. In Chapter 4 then we will introduce a new method of solution for "equilibria", using "Gibbsian expectation" instead of "mathematical expectation".

To get a feel for how to convert normal form games into decision reactions, let's look at a couple examples. We will look at the Prisoners Dilemma (PD) and Battle of the Sexes (BoS) games from Chapter 1, to demonstrate the ideas.

Example 2-1. The Prisoners Dilemma (PD) game representation. This example is adapted from Example 1-3, but here the game is represented not in normal form, but as a set of chemical reactions. In Chapter 4, the game will be solved not using mathematical expectation, but Gibbsian expectation principles.

The PD game was represented in normal form as

	b1 = quiet	b2 = tell
a1 = quiet	**0.5**, 0.5	**7.0**, 0.0
a2 = tell	**0.0**, 7.0	**5.0**, 5.0

Represent this game as a set of chemical reactors.

SOLUTION

The game is represented as a set of decision reactions in three separate reactors, which correspond to three agents required to make the final decision. Each reaction has a standard Gibbs energy of reaction ($\Delta g^0 / kT$); In Chapter 3 we will assess this value more quantitatively, although for now we will take its value to be simply the number of months of prison.

The first reactor is where Player A makes a decision.

$$(2\text{-}1) \quad \begin{aligned} a1 + b1 &\xleftrightarrow{A} A11, \quad \Delta g^0 / kT = +0.5 \\ a1 + b2 &\xleftrightarrow{A} A12, \quad \Delta g^0 / kT = +7.0 \\ a2 + b1 &\xleftrightarrow{A} A21, \quad \Delta g^0 / kT = +0.0 \\ a2 + b2 &\xleftrightarrow{A} A22, \quad \Delta g^0 / kT = +5.0 \end{aligned}$$

The reaction between a1 and b1 produces A11, and there is a catalyst and conditions shown, represented by "A". One might wonder why worry about a catalyst, if we are considering equilibrium reactions. After all, in first year chemistry, don't we learn that a catalyst does not affect the outcome of an equilibrium reaction? This is true ... as long as we wait long enough to reach equilibrium!

However, some reactions take a long time to reach equilibrium, and so we introduce catalysts to speed up the path toward equilibrium. Examples where this occur abound, including hydrocarbon isomerization, the water-gas shift reaction, esterification, and many more. The water-gas shift reaction even has low temperature and high temperature shift catalysts.[28] In effect, catalysts "turn on and off" many types of equilibrium reactions.

Similarly, Player B is represented as a reactor with the following reactions:

$$
(2\text{-}2) \quad
\begin{aligned}
a1 + b1 &\xleftrightarrow{\;B\;} B11, & \Delta g^0 / kT &= +0.5 \\
a1 + b2 &\xleftrightarrow{\;B\;} B12, & \Delta g^0 / kT &= +0.0 \\
a2 + b1 &\xleftrightarrow{\;B\;} B21, & \Delta g^0 / kT &= +7.0 \\
a2 + b2 &\xleftrightarrow{\;B\;} B22, & \Delta g^0 / kT &= +5.0
\end{aligned}
$$

Players A and B make their decisions in their own space. As a result, the amounts of A21 and B12 will have the most outcome, while A12 and B21 will have the least.

Now I consider another Player in the PD game, which is seldom considered explicitly: the Enforcer, who might be the judge or the police. We will call the Enforcer and its decision as Z.

$$
(2\text{-}3) \quad
\begin{aligned}
A11 + B11 &\xleftrightarrow{\;Z\;} Z11, & \Delta g^0 / kT &= +2.0 \\
A12 + B12 &\xleftrightarrow{\;Z\;} Z12, & \Delta g^0 / kT &= -1.0 \\
A21 + B21 &\xleftrightarrow{\;Z\;} Z21, & \Delta g^0 / kT &= -1.0 \\
A22 + B22 &\xleftrightarrow{\;Z\;} Z22, & \Delta g^0 / kT &= -3.0
\end{aligned}
$$

In the PD game, the Enforcer might truly not care which punishment is given, so that $\Delta g^0 / kT = 0$ for each reaction; however, the basis of the game is that Z wants A and B to tell on each other.

ASSUMPTIONS

Classical game theory contains many assumptions that are now treated explicitly from the start. Here are a few of them:

- Enforcer. The PD game usually assumes that enforcement is complete and unbiased. In practice, I suspect the jailer is manipulating the prisoners a bit to get both Players to the "tell" Possibility. The jailer is thus a type of game master.

- predisposition. The PD game usually assumes unbiased Players, who simply want to minimize their prison sentence. However, in this formulation, we must explicitly set the amounts or concentrations of a1, a2, b1, and b2 that each Player has at the start, and thus their Prior predispositions. For instance, if I were one of the Players, I might play the game differently if I had learned that I should "look out for #1", without considering it in this context.
- size. Sometimes Players do not have an equal say in the decision. As a result, the amount of a1 + a2 need not be the same as b1 + b2.
- rationality. Although "rational" players need not be selfish,[29] in classical game theory the word rational is usually equated to omniscient and selfish. Later in this chaper, I will show how we can explicitly consider altruism, vengeance, global thinking, and other FLAVORS.
- same game. Game theory usually assumes that both Players see the same set of Possibilities, Pains, and Perspectives; however, due to ignorance, bias, or deception, Player A likely sees these items as being different for B, than Player B does for himself or herself. Information asymmetry is often at the heart of wielding power in strategic situations.

We will solve this game in Chapter 4, after we consider Pains and Perspectives, and Gibbsian solutions.

Example 2-2. The Battle of the Sexes (BoS) game representation. This example is adapted from Example 1-4, but here the game is represented not in normal form, but as a set of decision reactions. In Chapter 4, the BoS game will be solved not using mathematical expectation, but Gibbsian expectation principles

---

The BoS game was represented in normal form as

	b1 = hike	b2 = movie
a1 = hike	***-3.5***, -2.3	***-0.2***, -0.4
a2 = movie	***+0.3***, +0.5	***-2.6***, -3.3

The representation in terms of decision reactions is similar to the previous example. The first reactor is where Player A makes a decision.

$$a1 + b1 \xleftrightarrow{\ A\ } A11, \quad \Delta g^0 / kT = -3.5$$

$$a1 + b2 \xleftrightarrow{\ A\ } A12, \quad \Delta g^0 / kT = -0.2$$

(2-4)

$$a2 + b1 \xleftrightarrow{\ A\ } A21, \quad \Delta g^0 / kT = +0.3$$

$$a2 + b2 \xleftrightarrow{\ A\ } A22, \quad \Delta g^0 / kT = -2.6$$

Similarly, Player B is represented as

$$a1 + b1 \xleftrightarrow{\ B\ } B11, \quad \Delta g^0 / kT = -2.3$$

$$a1 + b2 \xleftrightarrow{\ B\ } B12, \quad \Delta g^0 / kT = -0.4$$

(2-5)

$$a2 + b1 \xleftrightarrow{\ B\ } B21, \quad \Delta g^0 / kT = +0.5$$

$$a2 + b2 \xleftrightarrow{\ B\ } B22, \quad \Delta g^0 / kT = -3.3$$

If Players A and B make their decisions in their own space, we still need an Enforcer (Z). But in this case, let's have Z be neutral. Z acts using

$$A11 + B11 \xleftrightarrow{\ Z\ } Z11, \quad \Delta g^0 / kT = 0.0$$

$$A12 + B12 \xleftrightarrow{\ Z\ } Z12, \quad \Delta g^0 / kT = 0.0$$

(2-6)

$$A21 + B21 \xleftrightarrow{\ Z\ } Z21, \quad \Delta g^0 / kT = 0.0$$

$$A22 + B22 \xleftrightarrow{\ Z\ } Z22, \quad \Delta g^0 / kT = 0.0$$

We will solve this game in Chapter 4, when we consider Pains and Perspectives, and Gibbsian solutions.

In Example 2-1, I employ several ideas. First, the Possibilities are represented as short symbols, just as chemical elements are. In the period table of elements, helium is He, oxygen is O_2, and silver is Ag (from the Latin word for silver, argentum). Second, each Player is represented as a set of unit operations (Chapter 2), which include chemical reactors. In the chemical reactors, the Possibility knowlecules (a1, a2, b1, and b2) can react. In this chapter we will examine "equilibrium", although in a later chapter we will mention more about dynamics as well. Thus, one decision is made in A, and another decision is made in B. The outputs from these first two reactors requires that we explicitly consider a third reactor C, which includes a Player we might call the "enforcer". In Chapter 4 we will consider other ways the two Players can interact, through a Process Flow Diagram.

This representation using chemical reactors enables us to explicitly address assumptions normally made in game theory:

- enforcement. Usually game theory does not consider how the two separate decisions are enforced into a single decision. For example, in the Prisoners Dilemma (PD) game, we assume that after A and B choose, that the decision is automatic. However, the jailer must enforce the outcome, and can do so with various levels of vigor, depending on his or her desires, and capacity.

- lack of bias. Since we explicitly include the initial "concentrations" of Possibilities, we do not assume a neutral decision-maker to start with. Classical game theory uses Bayesian games to cover this possibility to some extent.

How do we assess the stoichiometry of reactions? If I have a chemical reaction A + B = 2C, where the equal sign means "in equilibrium with", I can rewrite this reaction as 0 = 2C – A – B, or

$$(2\text{-}7) \qquad 0 = \sum_i v_i M_i$$

where the M's represent the Possibilites (i.e., A, B, and C), and v_i represents the stoichiometric coefficients. It is seen that if 2 moles of C are produced, 1 mol of A is consumed, and 1 mol of B is consumed. Thus the stoichiometric coefficients are $v_A = -1$, $v_B = -1$, and $v_C = +2$. Given a change of moles given by dn_A, we can define an "extent of reaction" (ε) as

$$(2\text{-}8) \qquad \frac{dn_A}{v_A} = \frac{dn_B}{v_B} = \frac{dn_B}{v_B} = d\varepsilon$$

In general, when there are q Players, each with m_i Possibilities each, the number of reactions (r) will be

$$(2\text{-}9) \qquad r = (q+1)\prod_{i=1}^{q} m_i$$

The "1" added to the q is for an Enforcer, which can sometimes be omitted. Thus, for 2 Players with 2 choices, we have r = (2+1)(2)(2) = 12 reactions. For 2 Players with 3 each, r = (2+1)(3)(3) = 27 reactions. For 3 Players with 2 choices each, r = (3+1)(2)(2)(2) = 32 reactions. For 3 Players with 3 choices each, r = (3+1)(3)(3)(3) = 108 reactions. Obviously, the number (r) of reactions grows very quickly.

Perception functions lead to stoichiometric coefficients.

Humans tend to sense things roughly logarithmically. Sound loudness is given in decibels,[30] which are logarithmic. Sound tone is given by octaves, which are logarithmic. Similar relationships hold for weight, brightness, salary increases, temperature sensing, and other human perceptions. The relationship is described by the Weber-Fechner Law,[31] which describes the relationship between a stimulus and a perception:

$$(2\text{-}10) \qquad dp = k\frac{ds}{s}$$

If we define a threshold stimulus as s_0, then an integration gives an expression for a perception function

$$(2\text{-}11) \qquad p = k\ln\frac{s}{s_0}$$

Thus, the threshold stimulus has $p = 0$. We could of course establish a polynomial in the logarithm, if needed.

Let me examine the $p(s)$ "perception function" for money, in the following example. I will use a simple scale.

Example 2-3. Perception function for money.

Let's look at a clothing purchase of an average American worker in 2015. Say that a small amount is \$5 (e.g., a pair of socks), a medium amount is \$70 (e.g., a sweater), and a large amount is \$300 (e.g., a suit). The person can also estimate that a "zero amount" is <\$1, while an "infinite amount" is >$1000. What is the person's perception function for clothing?

SOLUTION. We will use a 5-point scale, with values ...

0 = infinitesimal = $0.50 (i.e., hardly worth thinking about)
1 = small = $5 (e.g., a pair of socks)
2 = medium = $70 (e.g., a sweater)
3 = large = $300 (e.g., a suit)
4 = infinite = $1000 (i.e., no way)

We might argue about the precise amounts, but let's see if a fit gives us something useful. Fitting these numbers gives

$$p = 0.50 \ln \frac{s}{0.67}$$

where s is the dollar amount, s_0 = $0.67, and the coefficient k = 0.50. The plot of the data appear here:

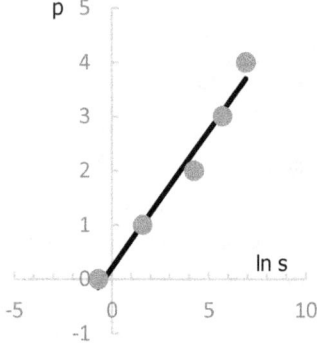

That is, if a pair of paints costs $35, the person would perceive this as p = 1.98, or roughly a medium expense.

Just for fun, let's examine what this perception function gives for the total wealth in the world! This value is estimated to be about $250 trillion. The p value from this perception function is 16.8.

I note that the use of a 5-point scale, or a 10 point scale, or other, is arbitrary at this point. Perhaps psychological studies reveal how many levels most people tend to hold in their minds at once.

Decisions can thus be written as chemical reactions, using "fundamental resources" or "elements" such as SMART resources:

S = stuff. Number of books, number of sweaters, number of cups.

M = discretionary money. Measured in $/week or other period.

A = active time. High energy time per week.

R = relationships. Number of friends, or a perceived quality.

T = discretionary time. Measured per week, for week-type decisions.

We might also imagine using Tr = trust, producing F = fear, producing H = hope, or gaining S = status. Hope would be a somewhat low energy knowlecule (i.e., probable), while fear would be a very low energy (i.e., probable) knowlecule.

Example 2-4. Perception function for time.

Let's look at an "average American person" in 2015, and how they view time with regard to major life events like high school, college, marriage, changing jobs, children. The median life span is about 80 years. What is the perceived length of time for a semester (16 weeks), when viewed on a lifespan?

SOLUTION. We will use a 5-point scale, with values ...

0 = infinitesimal = 1 day
1 = small = 1 month (e.g., a pair of socks)
2 = medium = 1 year (e.g., a sweater)
3 = large = 5 years
4 = infinite = 100 years

Fitting these numbers gives, with s in years,

$$p = 0.39 \ln \frac{s}{0.0040}$$

A plot is

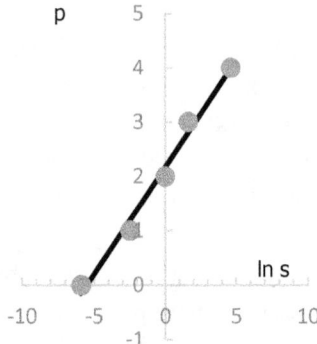

A semester is 16 weeks = 0.308 years, so p = 1.7, not quite a medium amount of time.

Just for fun, let's examine the age of the earth, which is estimated to be about 4.5 billion years. Now p = 10.8. Thus we see that according to our scale, an amount of about 10 is very large indeed.

Concentrations of reactants and products are also viewed through the perception functions. Thus, if we are buying apples, we establish a perception function for apples, and if we are purchasing 17 apples, we perceive that as some value, on the 0-6 perception scale.

Possibilities react to form new outcomes.

In chemical engineering, there are "phase transitions" and "chemical transitions". Phase transitions involve no new molecular species, but rather a change of phase. Solid water – ice – melts into liquid water. Liquid water evaporates into vapor water. Chemical transformations, on the other hand, bring about new molecular species. Methane and oxygen react to form carbon dioxide and water. In general, "separations" occur when two or more species are separated, without creating any new molecules. "Reactions" occur when multiple species interact, giving rise to new species.

Many evaluations are separation processes. If I am examining the price of a book from five different sellers, and making my selection based on price, that is an evaluation or separation. Creative activities, however, are reactive processes. When I was deciding whether to write this book, I had to assess whether it was favorable for me to convert my time, energy, money, and other resources into the decision to write a book. I knew that if I wrote the book, I would use up those resources.

Chemical reactions tend to move toward "low energy molecules", such as carbon dioxide (CO_2) or water (H_2O). High energy molecules like methane (CH_4) are reactive. In the context of creativity, some knowlecules like hope and especially fear are low energy (i.e., probable). If a decision involves fear as a potential outcome, that decision process tends to divert strongly toward the fear, unless tremendous energy is input to overcome it. That uses "willpower", which research shows is a finite resource.[32]

In every decision involving an outcome different from the inputs – that is, a thought reaction – there are a number of species that should be accounted for:

- decision outcomes. These are the outcomes anticipated readily, usually written in a normal form game.

- creative outcomes. These are less obvious, but especially when the pain of the decisions will be high, new ideas emerge creatively.

- fear, hope, regret. These are not always written, but they often drive decisions since they are high or low energy. The classic FUD recipe – fear, uncertainty, and doubt – are all forms of fear. Furthermore, fear can poison the catalysts which are mentioned later in this chapter, rendering them inoperative.

- corruption. Games almost always assume that people "play by the rules". That is of course not true in all cases. One should always consider illegal, corrupt, criminal, or violent outcomes as possibilities.[33] Unless the pain for these outcomes is high, they can emerge. Remember the motto: "People first abhor crime. If not checked, they accept it. And if that is not checked, they embrace it."

Stoichiometry is tracked using a SPICEY table.

Stoichiometry is one part of a large discipline called material balances or mass balances. In Chemical Engineering we use a stoichiometric table, sometimes called an ICE table (initial, change, equilibrium). Since I tend to see myself not as a cold person, but as a spicy person, I use a SPICEY table (SPecies, Initial, Change, Equilibrium, y the mole fraction).

Example 2-5. Chemical equilibrium for the formation of ammonia (NH_3) using the Haber-Bosch process.

We want to form ammonia (NH_3) starting from 1.6 mole nitrogen (N_2), 2.9 mole hydrogen (H_2), and 2.3 mole argon (Ar) gases. There is already 0.2 mole NH_3. Express the final gas mole fractions (y) of each species in a simple, rigorous way.

SOLUTION. The chemical reaction is

$$N_2 + 3H_2 = 2NH_3$$

This equation gives the stoichiometric coefficients as $v_{N2} = -1$, $v_{H2} = -3$, $v_{NH3} = +2$, and $v_{Ar} = 0$. We will use a single "extent of reaction" (ε), since the change in each species can be expressed by this one variable, through the chemical equation. The SPICEY table is

species	initial	change	equilibrium	y
N_2	1.6	$-\varepsilon$	$1.6 - \varepsilon$	$(1.6 - \varepsilon) / \Sigma$
H_2	2.9	-3ε	$2.9 - 3\varepsilon$	$(2.9 - 3\varepsilon) / \Sigma$
NH_3	0.2	$+2\varepsilon$	$0.2 + 2\varepsilon$	$(0.2 + 2\varepsilon) / \Sigma$
Ar	2.3	0	2.3	$2.3 / \Sigma$
			$\Sigma = 7.0 - 3\varepsilon$	

For any extent of reaction (ε), we can now find the final equilibrium amount, and the equilibrium mole fractions (y) using the table above. There are several important notes at this point:

- equilibrium. At this point the equilibrium value of ε is unknown. We will be able to solve for ε using the material in Chapter 3. We combine the material balance from this chapter, with the thermodynamics of Chapter 3 – in particular, the Gibbs energy change (ΔG) and the equilibrium constant (K) – to find the equilibrium extent (ε).
- rate. Equilibrium is not always reached quickly. The equilibrium state for the gasoline in a tank and the oxygen in the surrounding air is mostly carbon dioxide and water. However, gasoline can sit in a tank for many months without reaching this equilibrium.
- catalysts. Most industrial processes use a catalyst, which at the simplest level turn a reaction "on and off". For the Haber-Bosch reaction, one usually uses a magnetite or iron catalyst.

At this point we have expressed the equilibrium for a single reaction. We can do something similar for multiple reactions. I note that for multiple reactions, it can sometimes be difficult to identify the independent reactions, and there are matrix techniques for evaluating the number of truly independent reactions.

Example 2-6. Convert the Prisoners Dilemma (PD) game into a set of decision reactions, and then put into an SPICEY table. We have Players A, B, and the enforcer D.

In example 3-1 we wrote a set of reactions for the PD game. We will now add "enforcement reactions", which combine the separate decisions of A and B. These reactions are

$$a1 + b1 \xleftrightarrow{\;A\;} A11$$
$$a1 + b2 \xleftrightarrow{\;A\;} A12$$
$$a2 + b1 \xleftrightarrow{\;A\;} A21$$
$$a2 + b2 \xleftrightarrow{\;A\;} A22$$
$$a1 + b1 \xleftrightarrow{\;B\;} B11$$
$$a1 + b2 \xleftrightarrow{\;B\;} B12$$
$$a2 + b1 \xleftrightarrow{\;B\;} B21$$
$$a2 + b2 \xleftrightarrow{\;B\;} B22$$
$$A11 + B11 \xleftrightarrow{\;D\;} D11$$
$$A12 + B12 \xleftrightarrow{\;D\;} D12$$
$$A21 + B21 \xleftrightarrow{\;D\;} D21$$
$$A22 + B22 \xleftrightarrow{\;D\;} D22$$

Rewrite these in a stoichiometric SPICEY table. Assume that each Player has no *a priori* bias (i.e., learning) in their decision.

SOLUTION. We now have 12 extents of reaction (ε_i), and 16 species. We assume that each reaction is appropriately catalyzed to turn the reaction on. Since the Players have no bias, we will assume that the initial amounts $n_{a1} = n_{a2} = 0.5$ The table is then

species	initial (n_{0i})	change	equil (n_i)	x
a1	0.5	$-\varepsilon_1 - \varepsilon_2 -$	$0.5 -\varepsilon_1 - \varepsilon_2 -\varepsilon_5 - \varepsilon_6$	For all,
a2	0.5	$\varepsilon_5 - \varepsilon_6$	$0.5 -\varepsilon_3 - \varepsilon_4 -\varepsilon_7 - \varepsilon_8$	$x_i = n_i / \Sigma$
b1	0.5	$-\varepsilon_3 - \varepsilon_4 -$	$0.5 -\varepsilon_1 - \varepsilon_3 -\varepsilon_5 - \varepsilon_7$	
b2	0.5	$\varepsilon_7 - \varepsilon_8$	$0.5 -\varepsilon_2 - \varepsilon_4 -\varepsilon_6 - \varepsilon_8$	
A11	0.0	$-\varepsilon_1 - \varepsilon_3 -$	$+\varepsilon_1 - \varepsilon_9$	
A12	0.0	$\varepsilon_5 - \varepsilon_7$	$+\varepsilon_2 - \varepsilon_{10}$	
A21	0.0	$-\varepsilon_2 - \varepsilon_4 -$	$+\varepsilon_3 - \varepsilon_{11}$	
A22	0.0	$\varepsilon_6 - \varepsilon_8$	$+\varepsilon_4 - \varepsilon_{12}$	
B11	0.0	$+\varepsilon_1 - \varepsilon_9$	$+\varepsilon_5 - \varepsilon_9$	
B12	0.0	$+\varepsilon_2 - \varepsilon_{10}$	$+\varepsilon_6 - \varepsilon_{10}$	
B21	0.0	$+\varepsilon_3 - \varepsilon_{11}$	$+\varepsilon_7 - \varepsilon_{11}$	
B22	0.0	$+\varepsilon_4 - \varepsilon_{12}$	$+\varepsilon_8 - \varepsilon_{12}$	
D11	0.0	$+\varepsilon_5 - \varepsilon_9$	$+\varepsilon_9$	
D12	0.0	$+\varepsilon_6 - \varepsilon_{10}$	$+\varepsilon_{10}$	
D21	0.0	$+\varepsilon_7 - \varepsilon_{11}$	$+\varepsilon_{11}$	
D22	0.0	$+\varepsilon_8 - \varepsilon_{12}$	$+\varepsilon_{12}$	
		$+\varepsilon_9$		
		$+\varepsilon_{10}$		
		$+\varepsilon_{11}$		
		$+\varepsilon_{12}$		
			$\Sigma = 1.0 -\varepsilon_1 -\varepsilon_2 - \varepsilon_3$ $- \varepsilon_4 -\varepsilon_5 -\varepsilon_6 - \varepsilon_7 - \varepsilon_8$ $- \varepsilon_9 - \varepsilon_{10} - \varepsilon_{11} - \varepsilon_{12}$	

The meaning of these values is that the "initial" amounts n_{0i} are the starting concentrations of how the Players are inclined to play. The final quantities of Dij give the decision outcomes that result, when both Players have played. In the present example, I have assumed that all reactions occur in the same reactor. This is in fact an assumption, because as I show in Chapter 4, we can draw various dialogical processes to represent how the decision progresses.

In the table for Example 2-6, I have given that the initial concentrations of a_1 and the other species are all 0.5. Here is an additional advantage of representing the Possibilities as chemical species: We can in fact set a proper amount of each Possibility, based on the initial biases of the person! Thus, we could have 0.73 of a_1 and 0.27 of a_2, and indeed, there is no reason it has to add to unity, nor that the sum of b_1 and b_2 has to be the same as for Player A. This gives us full Bayesian flexibility, since our pre-biases can be estimated and used, rather than considering that we a priori see all Possibilities as being equal. The usual Bayesian approach in game theory is to update the pains, but here we can keep the pains as is, and adjust the a priori biases through the concentrations.

In producing the SPICEY table, I list all possibilities that register an initially finite concentration. If the initial concentration is 0, then in fact that species is not a Possibility at the start. This is especially important knowing that our memories are finite, and so the initial amount of items we can hold in our memory and process might be <10.[34,35]

The table also reveals the importance of our environment. Particular environments where we reside throughout a given day, might trigger the transport of small-concentration parts of our memory. Thus, given a certain environment, we might remember something that brings a new Possibility to bear. Likewise, we might forget a Possibility, in a different environment.

Chemistry can be expressed generally with matrices.

There is a more general way to represent the stoichiometry. Say we have we have s species and r reactions. I define a vector of initial concentrations:

$$(2\text{-}12) \qquad \mathbf{n}^0 = \left(n_1^0, n_2^0, ..., n_s^0 \right)$$

We have a matrix ($\mathbf{v}$) of stoichiometric coefficients for the various species (sp) that go into the reactions (rxn):

$$(2\text{-}13) \qquad V = \begin{pmatrix} v_{sp1}^{(rxn1)} & v_{sp2}^{(rxn1)} & \cdots \\ v_{sp1}^{(rxn2)} & v_{sp2}^{(rxn2)} & \cdots \\ \cdots & \cdots \\ & & & v_{sp\,s}^{(rxn\,r)} \end{pmatrix}$$

We have a vector of extent of reactions:

$$(2\text{-}14) \qquad \boldsymbol{\varepsilon} = \left(\varepsilon_1, \varepsilon_2, \ldots, \varepsilon_r \right)$$

Thus, the reaction proceeds according to

$$(2\text{-}15) \qquad \mathbf{n} = \mathbf{n}_0 + V \cdot \boldsymbol{\varepsilon}$$

We can use this more general representation to code the stoichiometry matrix more readily into a computer code.

There is one more important point with chemistry that I mention here: Decisions might involve a small number of knowlecules, as opposed to Avogadro's number of molecules for a typical chemical reaction. Professor James B. Rawlings at the University of Wisconsin-Madison studies this problem for chemical systems, and this could be a very fruitful area of study for decision reactions.

Catalysts act as on-off switches, in many FLAVORS.

Good catalysts increase yield, and also selectivity. They make reactions proceed more rapidly by decreasing the activation energy barrier. Catalysts increase the rate in both the forward and reverse directions, and so for equilibrium reactions, they can effectively turn the reaction "on and off". Catalysts are used for many equilibrium reactions, including the Haber-Bosch reaction for ammonia, the water-gas shift reaction, and steam reforming.[36] Catalysts can help assure that a particular reaction pathway becomes viable, but that side reactions are not viable. This is especially important when the catalyst is an enzyme, which can be especially selective. In fact, enzymes can be activated or inhibited, depending on particular molecules in the environment, so that the enzymes themselves do not always act the same.

In human decisions people can produce reactions by catalyzing various decision reactions. These can come in a variety of FLAVORS, seen from the Perspective of Player Q.

- Fair. Q wants to minimize differences among various participants in the final pain.
- Loyal. Q chooses to minimize the pain of a particular Player or group.
- Altruistic. Q chooses to minimize the pain of all Players except himself or herself.
- Vengeful. Q chooses to maximize the pain of a particular Player or group, without regard for himself or herself
- Overall. Q chooses the minimize the pain of all Players, including himself or herself. This is different from Altruistic.
- Rivalrous. Q chooses the maximize the pain of another Player or group, and minimize the pain for himself or herself.
- Selfish. Q chooses to minimize his or her own pain, without regard for others. "Selfish" is often – but not always – what is meant by "rational".

A person might want to emphasize any particular Player, and so use any of the FLAVORS above. For example, with my own daughters, I hope I would be less Selfish and more Overall, Fair, or Altruistic. We will see in Chapter 4 how to include the FLAVORS quantitatively, and the idea that they will turn on-off particular reactions.

It is also possible – perhaps probable – that each Player in a particular set of reactions has a different view of the game. One Player A will likely recognize a different set of Players, different Possibilities, different Pains, and have different Perspectives for the game, than does Player B or C.

Exercises

1. *chemical reaction.* Write the chemical reaction of glucose being oxidized to carbon dioxide and water.

2. *NO_2 reaction.* Write the SPICEY table for the gas phase reaction of nitrogen dioxide (NO_2) to dinitrogen tetroxide (N_2O_4), starting with 1.5 mol NO_2 and 0.3 mol N_2O_4.

3. *two reactions.* The following reactions take place:
    $C_2H_6 = C_2H_4 + H_2$
    $C_2H_6 + H_2 = 2CH_4$

    We start with only 1.0 mol ethane (C_2H_6). Write the SPICEY table that includes both reactions, and give an expression for the mole fraction of hydrogen (H_2) in terms of the extents of reaction.

4. *Battle of the Sexes game.* We have a Battle of the Sexes game, in which the game matrix is given below. Player A enjoys movies, and Player B enjoys sports. As usual, they want to be together, but also want to attend their favorite event. Write the

normal form game as a set of decision reactions. Include the reactants (Possibilities) and catalysts in your equations.

	b1 = movie	b2 = game
a1 = movie	**-3.0**, -2.0	**+0.1**, -0.4
a2 = game	**+1.3**, +1.5	**-1.8**, -3.2

5. *your time-energy-money (TEM) perception functions.* Use the Weber-Fechner law to find your own functions for time (discretionary hours per week), energy (energetic hours per week), and money (discretionary dollars per week). This law can be expressed as

$$p = k \ln \frac{s}{s_0}$$

Choose your TEM values per week, and use the Excel sheet to calculate your stoichiometric functions for these.

6. *buying a sweater with Henry.* You are going to buy a sweater, and you are going to take along our friend, Henry Louis Le Châtelier. Le Châtelier asks you to use your time and money perception functions to write a decision reaction for buying the sweater. While you are with Henry, you suddenly hear news that an appointment has been canceled, and so you have an extra hour of discretionary time. Then you find $30 in your coat pocket, which you didn't know you had – it is like extra money. State your values for p = 0 to 6, fit your time and money perception functions, write the decision reactions for buying a $70 sweater that takes 45 minutes to purchase, and state which will have a bigger impact, the extra $30, or the extra hour.

References

Covey, Stephen M.R.; Covey, Stephen R. *The Speed of Trust* (2008). Catalysts that account for trust and "overall" are usually more effective, including for selfish individuals, than selfish catalysts.

Denbigh, Kenneth. *The Principles of Chemical Equilibrium.* 4th ed. (1981). This has tremendous insights into phase and chemical thermodynamics. It is one of two thermo books I reach for first.

Dill, Ken A.; Bromberg, Sarina. *Molecular Driving Forces: Statistical Thermodynamics in Chemistry & Biology.* (2002). This is the other thermos book I reach for first!

Felder, R.M. and Rousseau, R.W. *Elementary Principles of Chemical Processes*, 3rd Edition (2005). The is the classic book on material balances for chemical engineers.

Fogler, H. Scott. *Elements of chemical reaction engineering.* 3rd ed. (1999). A standard reactor design book that emphasizes the fundamental balances: mass, energy, momentum.

Pauling, Linus. *General Chemistry* (1947).

Pauling, Linus. *The Nature of the Chemical Bond* (1939). This book contains many of the ideas that helped Pauling win the Nobel Prize in Chemistry in 1954.

3. Continuum. Pain potential, entropy, and equilibrium.

Nature has placed mankind under the governance of two sovereign masters, pain and pleasure. It is for them alone to point out what we ought to do, as well as to determine what we shall do. On the one hand the standard of right and wrong, on the other the chain of causes and effects, are fastened to their throne. They govern us in all we do, in all we say, in all we think.

– Jeremy Bentham[37]

Pain potentials quantify pain or utility.

One of the early utilitarians, Jeremy Bentham, promoted the "greatest happiness principle" for utility, meaning the rank of "pleasure" above "pain". This chapter bases what is usually called "utility" on "Pain". Negative pain is taken as pleasure. I make an analogy between a pain potential, and the usual chemical potential of chemical thermodynamics. Decision reactions tend toward the position of lowest pain, similar to how chemical reactions tend toward the position of lowest Gibbs free energy.

In making these analogies, I examine a rigorous way of quantifying utility, and therefore of quantifying decision making information processing. Whether this approach is applicable to human decisions is a matter for experimentation, but in any case, it is simply a different methodology from the usual game theory. More to the point, whether this approach is more descriptive of human behavior than classical game theory is a matter for experimentation; I am solving with a new approach, and making a different set of assumptions. In this book I will not seek to prove again the body of theory from thermodynamics, but in making the analogy, I call forth that theory.

The concept of a "potential energy" is well-known in physics. If I climb a set of stairs, I have increased my mechanical potential energy (J). If I place a wire across two battery terminals, I allow electrons to flow from high electric potential (i.e., voltage, V) to low. In the late 1800s, Josiah Willard Gibbs developed the idea of a "chemical potential". For chemical systems – and thus not including electrical, magnetic, gravitational, or other aspects – the Gibbs free energy (G, in J) is given by

$$(3\text{-}1) \qquad dG = -SdT + Vdp + \sum_i \mu_i dn_i$$

where S is the entropy (J/K), V is the volume of the system (m^3), p is the pressure (N/m^2 or J/m^3), μ_i is the chemical potential of species i (J/mol), and n_i is the number of moles of species i (mol). Thus, the chemical potential μ_i gives how much G increases, when a quantity dn_i is added to the system. For example, when a molecule of water leaves the liquid phase to enter the vapor phase, dn_w is negative for the liquid phase, and positive for the vapor phase. If the chemical potentials are favorable, the water molecule will move on average. If the system is at equilibrium, the molecule will not on average transfer between the gaseous and liquid states.

The chemical potential is defined as

(3-2) $$\mu_i = \mu_i^0 + RT \ln a_i$$

where μ_i^0 is the standard state chemical potential of component i, R is the gas constant (J/mol-K), T is the absolute temperature (K), and a_i is the activity of species i. For a general solution, including nonidealities, $a_i = \gamma_i x_i$, where x_i is the mole fraction of species i in solution, and γ_i is the activity coefficient of species i. For an ideal solution, all components have $\gamma_i = 1$, and so

(3-3) $$\mu_i = \mu_i^0 + RT \ln x_i$$

For gases, $a_i = \phi_i y_i p / p^0$, where y_i is the mole fraction of species i in the gas phase, p is the pressure, p^0 is the standard pressure (usually chosen to be 1 bar or 1 atm), and ϕ_i is the fugacity coefficient. For an idea gas, all $\phi_i = 1$, and so $a_i = y_i p / p^0$, and

(3-4) $$\mu_i = \mu_i^0 + RT \ln \frac{y_i p}{p^0}$$

The chemical potential allows us to include both the energetics of the first law of thermodynamics and the entropy of the second law. Chemical equilibrium is reached when the universe attains the maximum entropy, which indicates the "multiplicity", which in turn is the number of states in which a given energy can reside. In this book we will use both the energetic and entropic portions. Rather than listing arbitrary numbers in a normal form game matrix, we will list the "pain potentials" of the various decision reactions. I note that if all outcomes of a game are too painful, people might simply choose a different game to play, or produce new and unexpected Possibilities.

Gibbsian analysis can predict reaction equilibrium.

The usual and well-known case of a chemical reaction follows the minimization of the Gibbs free energy. This section gives relationships well-known in the Chemical Engineering community. We have already seen that for a chemical reaction,

$$(3\text{-}5) \qquad d\varepsilon = \frac{dn_i}{v_i}$$

so that $dn_i = v_i d\varepsilon$. Thus, at constant T and p,

$$(3\text{-}6) \qquad dG = \sum_i \mu_i dn_i = \sum_i \mu_i v_i d\varepsilon .$$

At equilibrium, the Gibbs energy is a minimum, so $\partial G / \partial n_i = 0$, and thus, $\partial G / \partial \varepsilon = 0$. As a result, the condition for the equilibrium of a chemical reaction occurs when

$$(3\text{-}7) \qquad \frac{\partial G}{\partial \varepsilon} = \sum_i \mu_i v_i = 0$$

This is a general condition of reaction equilibrium, as described in detail of Chapter 4 (Section 4.4) of Kenneth Denbigh.

Substituting (4-2) into (4-5) gives

$$(3\text{-}8) \qquad \sum_i \mu_i v_i = 0 = \sum_i \left(\mu_i^0 + RT \ln a_i \right) v_i$$

Defining a standard molar Gibbs free energy change

(3-9)
$$\Delta g^0 = \sum_i \mu_i^0 \nu_i$$

enables us to write

(3-10)
$$\Delta g^0 = -RT \sum_i \nu_i \ln a_i = -RT \sum_i \ln a_i^{\nu_i}$$

Now defining an equilibrium constant

(3-11)
$$K = \prod_i a_i^{\nu_i}$$

enables us to write

(3-12)
$$\Delta g^0 = -RT \sum_i \ln a_i^{\nu_i} = -RT \ln \prod_i a_i^{\nu_i} = -RT \ln K$$

So finally, we can write the well-known ideal expression

(3-13)
$$K = \exp\left(-\frac{\Delta g^0}{RT}\right) = \prod_i a_i^{\nu_i}$$

If we have an ideal liquid solution, then $a_i = x_i$ and so

(3-14)
$$K = \exp\left(-\frac{\Delta g^0}{RT}\right) = \prod_i x_i^{\nu_i} \quad \text{(liquid)}$$

If we have an ideal gas, then $a_i = y_i p / p^0$ and so

$$(3\text{-}15) \qquad K = \exp\left(-\frac{\Delta g^0}{RT}\right) = \prod_i y_i^{\nu_i} \times \left(\frac{p}{p^0}\right)^{\sum_i \nu_i}$$

Equation 3-11 contains both the enthalpic energy and the entropy of the system, and the equation has tremendous power. In the chemical sciences, Eq 3-11 allows us to add chemical reactants, allow them to equilibrate, and then measure the final product and reactant amounts. Thus, we can evaluate K, or equivalently, Δg^0. Now knowing the value of K (or Δg^0) for a given reaction, we can *predict* that by adding any combination of species, what will be the final equilibrium for the new case. Isn't that amazing?

Using the same equations, we can calculate ΔG for the reaction, for any given extent of reaction (ε). Let's do this for an ideal gas reaction.

$$(3\text{-}16) \qquad \Delta G = \int_0^\varepsilon \sum_i \mu_i \, dn_i = \int_0^\varepsilon \sum_i \mu_i \nu_i \, d\varepsilon' .$$

Substituting (3-4) for an idea gas and dividing through by RT gives

$$(3\text{-}17) \qquad \frac{\Delta G}{RT} = \varepsilon \frac{\Delta g^0}{RT} + \varepsilon \ln \frac{p}{p^0} \sum_i \nu_i + \int_0^\varepsilon \left(\sum_i \nu_i \ln y_i \right) d\varepsilon'$$

The y_i terms must be left in the integral, since they depend on ε. The first term on the right hand side (RHS) is due to the standard state Gibbs energy of reaction, and the second is due to the pressure. The third term – as Denbigh points out so brilliantly[38] on p 136 of his book – is thus a free energy of mixing term! That is, if a reaction proceeds and leaves less and less reactant, the tendency of the reactant to want to mix becomes stronger and stronger, which eventually stalls the reaction at equilibrium. *It is this term – due to entropy – that is missing in classical game theory.* And it is this entropic term which dramatically changes many of the conclusions reached in classical game theory, such as the presence of dominant solutions, spurious tipping points, and the presence of multiple Nash equilibria for games like BoS and Chicken.

Remember that (5-18) is not linear in extent of reaction (ε), since both the pressure (p) and the mole fractions (y_i) depend upon ε, unless the pressure is regulated in some other way. Let's follow up on Example 2-5, to see what we can say about equilibrium.

Example 3-1. Chemical equilibrium for the formation of ammonia (NH_3) using the Haber-Bosch process. This example follows from Example 4-5.

We want to form ammonia (NH_3) starting from 1.6 mole nitrogen (N_2), 2.9 mole hydrogen (H_2), and 2.3 mole argon (Ar) gases. There is already 0.2 mole NH_3. Predict the equilibrium amounts of each species at P = 3.5 bar and T = 50 C = 323 K. For this reaction, Δg^0 = -32,900 J/mol N_2 consumed, and Δh^0 = -92,220 J/mol N_2 consumed.

SOLUTION. We will use 4 steps toward solving an equilibrium problem, characterized by the acronym RATE.
 1. reaction. Write the species and the stoichiometry.
 2. accounting. Define the material balance acounting in a SPecies-initial-change-equilibrium-y mol fraction (SPICEY) table, and give the mole fractions and equilibrium constant value (K).
 3. thermodynamics. Find the Δg^0 for the reaction, and calculate the K using thermodynamics.
 4. extent of reaction. Find the extent(s) of reaction, by equating the K from the material balance and the thermodynamics.

From the previous example, we have the stoichiometric reaction as

$N_2 + 3H_2 = 2NH_3$

Next we will write the SPICEY stoichiometric table, to give the material balance.

species	initial	change	equilibrium	y
N_2	1.6	$-\varepsilon$	$1.6 - \varepsilon$	$(1.6 - \varepsilon) / \Sigma$
H_2	2.9	-3ε	$2.9 - 3\varepsilon$	$(2.9 - 3\varepsilon) / \Sigma$
NH_3	0.2	$+2\varepsilon$	$0.2 + 2\varepsilon$	$(0.2 + 2\varepsilon) / \Sigma$
Ar	2.3	0	2.3	$2.3 / \Sigma$
			$\Sigma = 7.0 - 2\varepsilon$	1.00

First we evaluate K at the standard state of T = 298 K and P = 1.0 bar, for which the standard state energies are given above. We will call this K_1.

$$K_1 = \frac{a_{NH3}^2}{a_{N2} a_{H2}^3} = \frac{y_{NH3}^2}{y_{N2} y_{H2}^3} \left(\frac{p}{p^0} \right)^{-2} = \exp\left(-\frac{\Delta g^0}{RT} \right) = 584{,}900$$

Next I make an approximation, which I expect to work fairly well given only a 50 C temperature change, that

$$\ln \frac{K_2}{K_1} \approx -\frac{\Delta h^0}{R} \left(\frac{1}{T_2} - \frac{1}{T_1} \right) = -2.880$$

where K_2 is the equilibrium constant at 50 C = 323 K. At 323 K, we thus find that $K_2 = 32{,}820$, and so we have

$$K_2 = \frac{y_{NH3}^2}{y_{N2} y_{H2}^3} \left(\frac{p}{p^0} \right)^{-2} = 32{,}820$$

This corresponds to a Δg^0 = -10,399 J/mol N_2 consumed at 323 K. Substituting from the ICE table, we have

$$K_2 = \frac{\left(\dfrac{0.2+2\varepsilon}{7.0-2\varepsilon}\right)^2}{\left(\dfrac{1.6-\varepsilon}{7.0-2\varepsilon}\right)\left(\dfrac{2.9-3\varepsilon}{7.0-2\varepsilon}\right)^3}\left(\frac{3.5}{1}\right)^{-2} = 32{,}820$$

$$\frac{\left(\dfrac{0.2+2\varepsilon}{7.0-2\varepsilon}\right)^2}{\left(\dfrac{1.6-\varepsilon}{7.0-2\varepsilon}\right)\left(\dfrac{2.9-3\varepsilon}{7.0-2\varepsilon}\right)^3} = 402{,}045$$

This is a single equation with a single unknown, and so we can solve this equation using Mathematica or Excel or other software. One finds that $\varepsilon = 0.942$ satisfies the equation.

Let's plot the ΔG for the reaction as a function of ε. We will keep the pressure at 3.5 bar.

$$\frac{\Delta G}{RT} = \varepsilon\frac{\Delta g^0}{RT} + \varepsilon\ln\frac{p}{p^0}\sum_i v_i + \int_0^\varepsilon\left(\sum_i v_i \ln y_i\right)d\varepsilon'$$

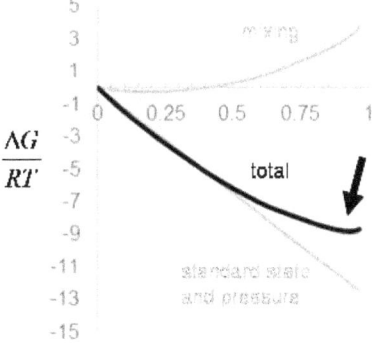

We see that the change in the Gibbs free energy due to the standard state energies plus the pressure – that is, the lower gray line –

continues downward in a linear fashion. But as we consume more and more of the H_2, we find that the H_2 "tries harder and harder" to mix back in, and thus the mixing part of the free energy goes up steeper and steeper. These combine to give a minimum Gibbs free energy at $\varepsilon = 0.942$, as found before, and as shown by the arrow in the plot.

The entropy of mixing almost always prevents a reaction from having an extent of 0 or 1. The exception might be the case like combustion, where the change in Gibbs energy is so high – 100s of RT per mol of methane burned with oxygen for instance – that due to the finite number of molecules, it is unlikely that even a single molecule of methane remains at the end.

We will use these reaction equilibrium concepts in Chapter 4, when we expand them to examine games.

Decision reactions can produce demand curves.

In high school chemistry or college first year chemistry, we evaluate Δg's for reactions using tables in the back of the book. No such tables exist for decisions yet ☺ So how will we find the Δg's for a particular decision reaction? Actually, we will use the same technique that is commonly used in chemistry: We will evaluate – actually we will estimate – the final equilibrium for an individual reaction, and then back calculate the Gibbs values from $K = \exp(-\Delta g^0 / RT)$. This is similar to how the Δg's for reactions were found for chemical systems. Let's look at an example of buying a sweater, for an example.

Example 3-3. Demand curve for a sweater.

Mary is going out to buy a sweater. She has put aside a $200 clothes budget for her wardrobe, has spent none of it yet. Currently, she has 2 nice sweaters that she likes. She goes to the store with $150 in her pocket, and thinks, "If I see a sweater I like for $70, I'll buy it. If it's more, I'll see how it goes." She has nothing else on her mind when she goes – no distractions. When she goes to the store, she sees a sweater she likes for $75. But she starts to wonder what her own demand curve is, for various prices, say from $30 to $150. And how does this depend on whether she has 2 sweaters, or 5, or 0 distractions, or 3?

SOLUTION.
We will use the 4 RATE steps toward solving this problem, as we did for the reaction equilibrium. First we must establish the reaction.

$$v_M M = v_G G$$

This reaction says that $75 of money (M) will be in equilibrium with one sweater (the good, G). We will ignore other required resources, such as the time required to peruse and make the purchase. For convenience, we will use the perception function for money from Example 2-3:

$$p = 0.50 \ln \frac{s}{0.67}$$

We will also fit for the sweater function. Let's use the following perceived amounts for p(s):

0 = 0.5 sweaters (i.e., less than 1 sweater)
1 = 2 sweaters
2 = 5 sweaters
3 = 12 sweaters
4 = 20 sweaters

Fitting these gives the following equation and plot.

$$p = 1.06 \ln \frac{s}{0.63}$$

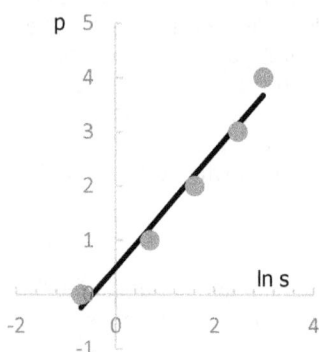

We see that a parabola might be a better fit, but a line is not terrible. Using the functions, we find that $75 is p = 2.36 and 1 sweater is p = 0.49., so

$$2.36M = 0.49G$$

The next step is to establish the material balance <u>accounting</u>. We will use a SPICEY table, and ignore any distractions for now. I will write the SPICEY table with both numerical values, and perceived values, for the $75 price. We can of course have a similar table for any price, number of distractions, or initial number of sweaters.

species	initial	change	equilibrium (n)	y
money	150	-75ε	$150 - 75\varepsilon$	
M	2.71		$0.50 \ln \dfrac{150 - 75\varepsilon}{0.67}$	n_M / Σ
sweaters	2	$+\varepsilon$	$2 + \varepsilon$	
G	1.22		$1.06 \ln \dfrac{2 + \varepsilon}{0.63}$	n_G / Σ
D	0		0	
			$\Sigma = n_M + n_G + n_D$	1.00

In the table I have included "distractors", which at first we will let be 0. The equilibrium constant is thus written as

$$K = \frac{y_G'^{0.49}}{y_M'^{2.36}} = \frac{\left(\dfrac{n_G}{\Sigma}\right)^{0.49}}{\left(\dfrac{n_M}{\Sigma}\right)^{2.36}} = f(\varepsilon)$$

The exponents depend upon the stoichiometric coefficients, which depend upon the initial amount of money and sweaters, as well as the money and sweater perception functions.

    If we knew the thermodynamics (i.e., the Δg^0 given Mary's 2 sweaters, \$150, and money and sweater perception functions), we could use it to calculate the extent of reaction at once. In a usual chemical reaction, we could find the Δg^0 values in tables or using the right computer software.

    However, we must first find our own <u>thermodynamics</u>, which is the third RATE step! To do this, we use the condition that Mary set: If she had \$150, and the sweater cost \$70, she would buy it. That is, $\varepsilon = 1$ at this point, and this gives us an equilibrium at a known point. We thus use the following SPICEY table:

species	initial	change	equilibrium (n)	y
money	150	-70ε	$150 - 70\varepsilon$	
M	2.71		$0.50\ln\dfrac{150-70\varepsilon}{0.67}$	n_M / Σ
sweaters	2	$+\varepsilon$	$2 + \varepsilon$	
G	1.22		$1.06\ln\dfrac{2+\varepsilon}{0.63}$	n_G / Σ
D	0		0	
			$\Sigma = n_M + n_G + n_D$	1.00

Using the value of $\varepsilon = 1$ sweater, we find that $n_M = 2.39$ and $n_G = 1.65$, so $\Sigma = 4.04$, and thus $y_M = 0.59$ and $y_G = 0.41$. Now we can evaluate that K = 2.24, which gives $\Delta g^0 / RT = -\ln K = -0.81$. Thus, it is slightly spontaneous that Mary will buy a sweater, but it depends upon the price. Now that we have this value, we can revert to the original table and solve our problem for all costs, not just $75.

We can now finalize the problem, calculating ε for any given initial amount of money, any given initial amount of sweaters, any given initial amount of distractors, and any given price. Here is the result for the demand curves:

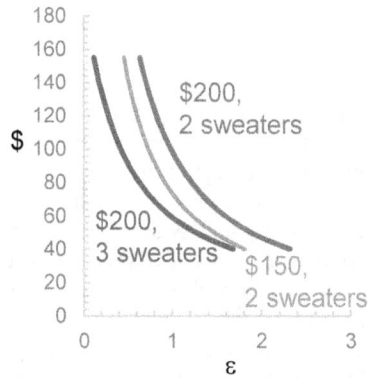

Note that as the price increases, the number purchased (ε) decreases. Also, as the initial amount of money drops from $200 to $150, ε decreases. And as the initial number of sweaters increases from 2 to 3, the tendency drops for purchasing an additional sweater at the same price. All of these are expected outcomes, and here we quantify them. We were able to draw these curves – and any others – using limited data.

We can solve for supply curves in a similar way. Later we will explore the idea that games can be solved as supply and demand concepts in a market.

If the Pains for all outcomes of a game lead to great Pain for either or both of the Players, an Enforcer will be required to make sure the game is played. That is, for the game to proceed spontaneously, the Players all need to feel that they can win at the game – and remember, a game need not be zero-sum – or that each Player at least has a reasonable probability of winning at the game. Otherwise, one or both of the Players simply avoid the game almost always, with no Enforcer present.

The knowlecular view enables kinetics to be assessed.

Having the knowlecular view of decision making, we can examine reaction kinetics and transport in decision systems. These include …

- creativity. As the Veloz et al article discusses,[23] we can use the socio-chemical metaphor. Based on the concepts in this book, we can also now add the energetics.
- temperature. I have more to discover about the precise meaning of temperature. It will open possibilities for using the Arrhenius equation. My current hypothesis is that temperature might be "measurable" in the same way humans "measure" temperature using thermoreceptor[39] systems to sense "hot", "warm", "comfortable", "cold". We could probably take numerous experiences, calibrate them on a scale of emotional level – perhaps with a Weber-Fechner representation, and start there. More precise measurements could involve actual physiological measurements of emotion.
- pressure. I likewise have more to discover about the precise meaning of pressure.
- diffusion. The diffusion of information has been written about in the book by Rogers.[40] The results have the possibility of being quantified using a PC approach.
- reaction kinetics. In this book I emphasize decision reaction equilibrium, but the viewpoint opens the possibility for examining kinetics. Having the perspective

of seeing decisions as sets of chemical reactions, we are not restricted to equilibrium. We can begin to use kinetic expressions for the rate, which can then be inserted into a material balance. The rate expressions are constitutive expressions that we can insert into the material balances. We use other constitutive relations for heat transfer (Fourier's Law), fluid flow (Newton's law of viscosity), elasticity (Hooke's law), and electrical conductivity (Ohm's law).

- bounded rationality. We need not reach equilibrium in all situations. Sometimes enough of the product can form at a particular temperature, to produce a decision. Such a decision might be satisfactorily sufficient (satisficing, in the words of Herb Simon).

It has been proposed previously to think of decision making processes as mixtures of thoughts. The "Garbage Can Model"[41] essentially describes a Continuous Sitrred Tank Reaction (CSTR), in which for example agendas and alternatives are mixed, until at a critical catalytic moment, a public policy might arise.[42]

Information has both entropic and enthalpic aspects.

This section contains some notes about "information", from a knowlecular viewpoint rather than the classical information theory viewpoint. Claude Shannon's classic work on information theory has formed the basis for digital and analog communication for decades. A core idea is that if there are n possible messages that can be sent, labeled i = 1 to n, and each message has a probability (p_i) the information (H) is given by

$$(3\text{-}18) \qquad H = -\sum_i p_i \log_2 p_i$$

The base 2 logarithm was used by Shannon, and gives "bits". I could use base 10 logs and get "digits", or base e logs and get "nats", or any other log. In this book I will generally express information in nats. The usual examples to check are the coin flip, or the dice throw, or letters in the alphabet.

Example 3-4. Information required to know a coin flip.

We flip a coin repeatedly. A fair coin has a probability $p_h = 0.50$ of getting a "heads", and $p_t = 1 - p_h = 0.50$ of getting tails. Let's say that we have a coin of various values of p_h. How much information (in bits) is needed to know the outcome of a flip?

SOLUTION
Inserting the results into our expression for information (H) enables us to write $H(p_H)$ and so have the entire curve.

$$H = -\sum_i p_i \log_2 p_i = -p_H \frac{\ln p_H}{\ln 2} - (1 - p_H)\frac{\ln(1 - p_H)}{\ln 2}$$

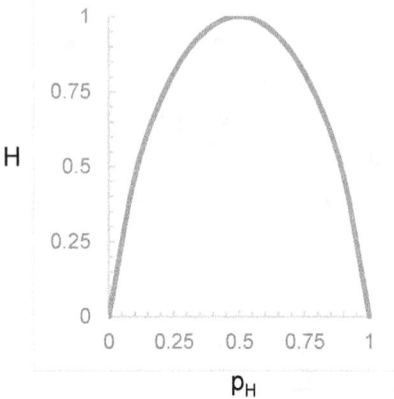

If we know for sure that the coin will land heads (i.e., p_h = 1.0), then no information is required to know the outcome of the flip. A similar result occurs if the coin is biased to always land tails. A "fair coin" requires 1 bit of information to know the outcome. If the coin has a 30% chance of landing tails (i.e., p_H = 0.70), then H = 0.881. That is to say, less information is needed to know the outcome of the flip. This becomes evident if we flip 1000 times.

The challenge that Shannon addressed was how to transmit information across a noisy channel. Previous researchers worked to reduce the noise in the channel. Shannon's approach was more clever: He showed that by coding the information in blocks, one could transfer the information from a source to a receiver, up to the capacity of the channel, with an arbitrarily small probability of error.

In Shannon's original work, he examined letters of the alphabet. He noted that in English, statistics reveal that an "e" is the most common letter (12.7%), while a "z" is the least common (0.074%). He was then able to calculate the information of a letter.

Example 3-5. Information in an alphabet character.

How much information is present in an alphabet letter, if all are equally likely, versus if they are distributed by the usual English frequency, given here?

a	8.17%	j	0.15%	s	6.33%
b	1.49%	k	0.77%	t	9.06%
c	2.78%	l	4.03%	u	2.76%
d	4.25%	m	2.41%	v	0.98%
e	12.70%	n	6.75%	w	2.36%
f	2.23%	o	7.51%	x	0.15%
g	2.02%	p	1.93%	y	1.97%
h	6.09%	q	0.10%	z	0.07%
i	6.97%	r	5.99%		

SOLUTION
First, we examine the case if all letters have equal probability. Inserting the results into our expression for information (H) gives

$$H = -\sum_{i=1}^{26} p_i \ln p_i = -\sum_{i=1}^{26} \frac{1}{26} \ln \frac{1}{26} = 3.258$$

If on the other hand we use the actual probabilities, the sum is

$$H = -\sum_{i=1}^{26} p_i \ln p_i = -(0.0817 \ln 0.0817 + 0.0149 \ln 0.0149...) = 2.894$$

Shannon showed schemes by which we could transfer this information across a channel, to take advantage of the fact that there is actually less information than an equal probability distribution. He also extended from letters, to words, to phrases. Coding mechanisms like the Huffman code are used to perform the actual encoding.

Shannon started with the fact that the probabilities are known. However, with our "knowlecular view" of letters, we might imagine that we have a continuous stream of letters flowing, and that certain letters "stick" or adhere to a surface more often than others. Mixing is spontaneous for ideal gases. But if I mix toluene and water 50-50, un-mixing is spontaneous! It depends on the chemical potentials as a function of the concentrations. Does air mix with molten Hg? Very little! In effect, Shannon assumes ideal mixing, and ignores energetics.

This allows "disinformation" to play an important role. For example, if I have 95 legitimate papers about smoking, and 5 papers give contradictory results – even if erroneous – then those 5 papers will carry disproportionate weight![43]

For letters, we find that some letters occur more often than others. What if we considered this to be a binding reaction? That is, an "e" binds much more strongly than a "z". How much stronger?

I'll use an adsorption reaction to characterize the binding energy of the letter "e" to a substrate (S):

$$e + S = eS$$

I define an equilibrium constant for binding

$$K_e = \frac{x_{eS}}{x_e x_S} = \frac{0.127}{(0.0385)(1)} = 3.302$$

The $0.0385 = 1/26$, and I let the substrate concentration start at 1. I can now calculate a binding energy

$$\frac{\Delta g_e^0}{RT} = -\ln K_e = -1.19$$

On the other hand, for "z" with a frequency of 0.00074, $K_z = 0.0192$ and

$$\frac{\Delta g_z^0}{RT} = -\ln K_z = +3.95$$

I now hypothesize the following: The frequency of the letters is not simply "given", but rather there is an "attraction" for certain letters. And if rather than being given 12,700 e's and 7.4 z's, we are given 5,000 e's and 5,000 z's, we can calculate the proportion of e's and z's that would be recognized. This is a different way of looking at frequency from the usual cryptology methods. Shannon saw information with mixing, but why didn't he include a term like Δg^0? In some sense classic information theory treats decisions as "happening at infinite temperature", where entropy dominates, while game theory treats decisions as "happening at zero temperature", where entropy plays no role.

Now I propose a different problem: Addressing a person's attention. I once attended a seminar at Carnegie Mellon University, while I was there as a PhD student, in which one of the speakers was Nobel Prize winner Herb Simon. He said, "The problem I find is very seldom that I cannot get enough information, but rather that there is too much information, and I have difficulty selecting what is relevant." That statement packs enormous insight.

The idea is that amidst the enormous stream of information that comes at us in life, we are educated to notice particular pieces of information. We notice when someone yells our name in a crowd; we notice when we hear our favorite song amidst a thousand competing sounds; we notice beautiful faces. There is a Gibbs energy of binding.

One challenge is that this enables deceptive people to poison our senses! Someone can produce a "sticky" knowlecule that binds, even though it does not contribute to the value of a decision! These could be knowlecules of 4 F's that are well known: Given some sudden situations, people throughout history have been known to fight, flee, freeze, or f__k! These can be enormously destructive knowlecules if they adsorb to our decision making apparatus! Deceptive knowlecules act as impurities that are attracted to our decision making apparatus.   If someone designs these sticky decision knowlecules well, they can deceive us and distort our decisions.   To remove these deceptive knowlecules can require enormous energy to engage the separation.

Mixture separations (evaluations) arise from pain potentials.

Starting from pure substances, the mixing entropy is given by

$$(3\text{-}19) \qquad \Delta S = -n \sum_i x_i \ln x_i$$

If a small amount of contaminant is added, there is a large ΔS of mixing, and as a result, a large negative ΔG of mixing. To un-mix a contaminant require energy. In chemical engineering, we usually use stage operations like distillation to separate mixtures. This is a process that could be examined further with information, and the addition of deceptive or non sequitur information. The key is that for any finite number of stages, one can never fully separate out a mixture! This is due not only to the entropy of mixing, but also the fact that staged operations must include a material balance.

I note that bringing in small concentrations of knowlecules can happen not only by deception or choice flooding, but also by triggers in the environment that recall "long lost knowlecules" remembered from previous times. That is, our environment has triggers that can cause a recall or forgetting of information.

Our memories, in contrast to many molecular systems, have only a finite capacity. Miller wrote his article indicating that we can hold 7 plus or minus 2 chunks. If a game has too much info, we have bounded rationality and cannot compute it. Something has to leave our memory, meaning that reaction does not occur. Usually we forget the small pains.

It is interesting to note that in using Eq 3-19, Shannon assumes all entropy, with no enthalpic contribution. In a similar way, game theory usually assumes all enthalpic contribution with no entropy. The Gibbsian games approach uses both.

Exercises

1. *NO_2 reaction.* Write the SPICEY table for the gas phase reaction of nitrogen dioxide (NO_2) to dinitrogen tetroxide (N_2O_4), starting with 1.5 mol NO_2 and 0.3 mol N_2O_4.

2. ammonia reaction. This problem follows from Example 3-1. We want to form ammonia (NH_3) starting from 3.1 mole nitrogen (N_2), 10.2 mole hydrogen (H_2), and 1.7 mole argon (Ar) gases. There is already 0.4 mole NH_3. Predict the equilibrium amounts of each species at P = 1.5 bar and T = 25 C = 298 K. For this reaction, Δg^0 = -32,900 J/mol N_2 consumed, and Δh^0 = -92,220 J/mol N_2 consumed.

3. *two reactions.* The following reactions take place:
   $C_2H_6 = C_2H_4 + H_2$
   $C_2H_6 + H_2 = 2CH_4$

Find the Δg^0's for the two reactions, and then solve for the final amounts, if we start with 3.0 mol of ethane, and none of the other species. Let T = 298 K and p = 1.7 bar.

4. *demand curve.* Reproduce Example 3-3. Include the Excel plots and full calculations.

5. *information for dice roll.* We roll two dice, which give a sum from 2 to 12. How much information is required to know the sum, compared with how much information is required to know the value of the two dice separately? Use units of nats.

6. *information theory.* We start with an NCAA March Madness bracket, with 68 teams. The details of the tournament can be found on Wiki. How much information is required to specify every win and loss, if all the teams have a 50-50 shot of winning in every game? How much information is required if in each game, there is a 90-10 probability of winning? The total number of games = 4+ 32 + 16 + 8 + 4 + 2 + 1 = 67.

References

Denbigh, Kenneth. *The Principles of Chemical Equilibrium.* 4th ed. (1981). This has tremendous insights into phase and chemical thermodynamics. It is one of two thermo books I reach for first.

Dill, Ken A.; Bromberg, Sarina. *Molecular Driving Forces: Statistical Thermodynamics in Chemistry & Biology.* (2002). This is the other thermos book I reach for first!

Gibbs, J. Willard. *The Scientific Papers of J. Willard Gibbs, Vol. 1: Thermodynamics* (1993). He did the work in the 1870s, and handed us the concepts of chemical potential, free energy (including entropy), and many others. My book could not exist without the work of Gibbs. I am in awe of him.

Shannon, Claude E.; Weaver, Warren. *The Mathematical Theory of Communication* (1971). This book is a near reproduction of Shannon's original two 1948 articles in the *Bell System Technical Journal.*

Simon, Herbert A. *Administrative Behavior*, 4th ed. (1997; 1st ed. in 1947). This book introduces the notion of "satisficing", meaning that there is no such thing as "economic man" (Homo Economicus).

4. Process. Connected unit operations.

All important unit operations have much in common, and if the underlying principles upon which the rational design and operation of basic types of engineering equipment depend are understood, their successful adaptation to manufacturing processes becomes a matter of good management rather than of good fortune. – Preface, Principles of Chemical Engineering[44]

People fight to be included in decision-making processes.

Those who have privilege, wealth, and power are in a position of relative advantage. They are connected to other people, who enhance their own power. That does not necessarily make them "bad people", just "more powerful people". Power determines who sets the agenda, who sets the rules of the engagement, and who sets the resources and therefore the tenacity in the battle.[45]

Those without connections to powerful people can fight to gain that power. A wizard of techniques for gaining such connections – sometimes praised, and sometimes vilified – was Saul Alinsky. In his book *Rules for Radicals*,[46] he wrote the following rules for gaining power in a situation:

1. "Power is not only what you have, but what the enemy thinks you have." Power is derived from 2 main sources – money and people. "Have-Nots" must build power from flesh and blood.

2. "Never go outside the expertise of your people." It results in confusion, fear and retreat. Feeling secure adds to the backbone of anyone.

3. "Whenever possible, go outside the expertise of the enemy." Look for ways to increase insecurity, anxiety and uncertainty.

4. "Make the enemy live up to its own book of rules." If the rule is that every letter gets a reply, send 30,000 letters. You can kill them with this because no one can possibly obey all of their own rules.

5. "Ridicule is man's most potent weapon." There is no defense. It's irrational. It's infuriating. It also works as a key pressure point to force the enemy into concessions.

6. "A good tactic is one your people enjoy." They'll keep doing it without urging and come back to do more. They're doing their thing, and will even suggest better ones.

7. "A tactic that drags on too long becomes a drag." Don't become old news.

8. "Keep the pressure on. Never let up." Keep trying new things to keep the opposition off balance. As the opposition masters one approach, hit them from the flank with something new.

9. "The threat is usually more terrifying than the thing itself." Imagination and ego can dream up many more consequences than any activist.

10. "The major premise for tactics is the development of operations that will maintain a constant pressure upon the opposition." It is this unceasing pressure that results in the reactions from the opposition that are essential for the success of the campaign.

11. "If you push a negative hard enough, it will push through and become a positive." Violence from the other side can win the

public to your side because the public sympathizes with the underdog.

12. "The price of a successful attack is a constructive alternative." Never let the enemy score points because you're caught without a solution to the problem.

13. "Pick the target, freeze it, personalize it, and polarize it." Cut off the support network and isolate the target from sympathy. Go after people and not institutions; people hurt faster than institutions.

All of these pieces point to one thing: a fight! Those who are at a disadvantage will either roll over, or fight.

PC analyzes the "unit operations" of people or agents.

Analyses of human decision making often begin with "people". Here we are going to treat people as "assemblies of unit operations". Chemical Engineering uses numerous chemical operations, which have a rough translation to decision making (Table 2-1).

Each Player or person or agent contains aspects of all these operations. Rather than list "Darrell" as a Player, in this book I list out the unit operations that Darrell brings to a decision making process. If he is considering a creative solution, I use a reactor; if he is evaluating two choices, I use a separator.

Teams therefore are not viewed as a collection of "people", but as a collection of unit operations. I might be creative in some aspects of my life, but when I work with other people on a given project, my creative side might lie mostly dormant, whereas in other situations it might be quite important. Thus, in this book I see a person bringing all his or her unit operations into the mix, and using them to various degrees.

Table 4-1. Unit operations in Chemical Engineering, and their similarity in human decision making.

unit op	CH E function	decision function
reactor	chemical reactions, transforming chemical A to chemical B.	decision making, transforming information inputs to decision outputs. Also, creativity.
separator	separation of an input into more pure outputs	evaluation of ideas
energy exchanger	heat, cooling, boiling, condensing	encouraging or cooling off
storage tank	storage, mixing	memory
pump	moving fluids forward	moving ideas forward

In establishing the Players, I look first for specific names of people – not simply institutions or companies or groups – and then establish the unit operations that each named person brings to the process. In many or most cases, we can think of PFDs as dynamic networks of agents. Even a single person can have multiple unit operations, and thus unit operations are fundamental pieces of a PFD.

Dialogical and decision processes can be drawn as a PFD.

Chemical processes consist of reactors, separators, mixing tanks, energy exchangers, pumps, and other unit operations. These unit operations are connected by pipes, which typically have valves at some point along their length. Most chemical processes use fluids, either liquid or gas, although many also use solid granular flows. Chemical processes are represented by process flow diagrams (PFD), with an example shown in Figure 2-1.

Flow diagrams are a great way to represent processes, whether we are examining chemical processes, algorithms, or other processes. The core idea here is that entire decision processes or dialogical processes can be represented by PFDs. This enables the more explicit design of dialogues, to allow for better outcomes, potentially for all parties. Dialogical processes consist of ways of knowing, of learning, of imagining, of analyzing.

Decision processes have "information flows". The information flows through the entire network of unit ops … and at each unit op, the information has the possibility of being degraded, purified, or transformed. These information flows are thus subject to the "noise" than Claude Shannon studied.

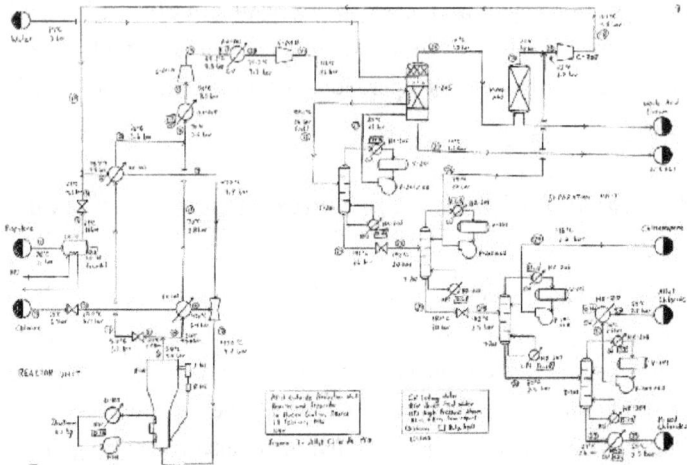

Figure 4-1. A process flow diagram for the production of allyl chloride (1992). This image is from my senior project at West Virginia University, and is a simplified version of an already simple PFD. The writing is too small to see, but the general diagram is visible, showing unit operations connect by pipes.

Network theory has a rich history in chemical engineering. Chemical reaction networks have been studied since at least the time of Rutherford Aris in the 1960s. Heat exchanger networks have been studied for decades. But PFDs have been the key design tool for creating flow networks. Piping and instrumentation diagrams (P&IDs) are also critical, but the PFDs show the major unit operations succinctly. The first step in understanding a decision process is to draw the PFD, so that we have an overview of the entire process.

Recycles in the networks enable Players to have repeated games. We purge information to enable us to consider problems. We have mixing of information. Some processes are in parallel, others in series. Some processes are successive, and some are simultaneous. Sequential decisions are made explicit using a PFD.

Example 4-1. PFD for purchasing a college text. This example follows from Example 1-5.

___

There were two Players or agents described: Jill and the bookstore manager. We will also include the professor, who has been charged with teaching a particular course, CH E 301 Thermodynamics. Let's say that the bookstore manager aims to produce as much net revenue as possible. Draw a PFD of the decision process.

SOLUTION. There are various versions of the PFD we could draw. We will draw a PFD from the perspective of the bookstore manager. There will be several decisions, some simple evaluations, and some decision transformations.

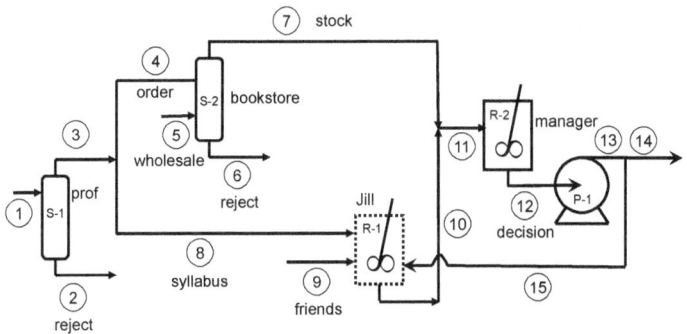

The streams are numbered 1-12 (here in circles, for easy visibility). The unit operations are numbered ...

S-1. The professor makes an evaluation about the book, given that the course will be taught (stream 1). This evaluation is based on a menu of books from memory or for instance Amazon.com (not shown). She selects a book (stream 3), and rejects the others (stream 2). In this PFD, there is no input from the manager, although this PFD might reveal that opportunity. The book decision is conveyed by the order (stream 4) to the manager, and to the student by the syllabus (stream 8).

S-2. The manager – with her "bookstore cap" on, makes an evaluation about the book, based on the order from the professor, and the input of the prices from various wholesalers (stream 5). This decision is based only on price, and so there is no decision reaction that happens. The decision is either to stock the book (stream 7) or reject (stream 6).

R-1. Jill has a decision to make, and it is based on the price of the book (stream 15), and the input from what her friends are doing (stream 9). This unit op is written using dotted lines, since the manager must make an educated guess about what Jill will decide (stream 10).

R-2. The manager must make a decision about the price (stream 12). The inputs are Jill's decision (stream 10), and the fact that the book is stocked (stream 7).

P-1. The decision is pumped out (stream 13), with part going back to the "assumed Jill" (stream 15), and part going to the bookshelf sticker price (stream 14).

This diagram might seem complicated for a decision like this. But it suggests several useful possibilities:

- faculty-bookstore communication. The bookstore doesn't want to get stuck with books. They need to develop an easy-to-submit communication which indicates how much latitude exists in choosing the book. This is especially important for large classes.
- required or recommended. The degree of "required" must also be communicated well.
- wholesale price. If the student is willing to pay less than the asking price, perhaps a different arrangement can be made to "fractionally purchase" the book, at the level needed.

Communications like these can facilitate proper strategic decisions that will provide improved value to students, the bookstore, and to publishers and wholesalers.

Each Player has an approximation of the PFD.

Each Player has only an approximation of the actual decision making PFD. There are many reasons for this:

- bounded rationality. Each Player has the ability to gather only a limited amount of information, and then each Player has a limited ability to process this information. Thus, the amount of "rationality" that a Player can have is bounded, as Herb Simon has stated. Herbert Gintis pointed out the "bounds of reason"[1] as being due to our social epistemology.

- information asymmetry. In many cases, one Player has a higher quantity or quality of information than the other, lead to information asymmetry.

- deception and trust. Since Players have only a finite capacity for learning information, if other Players contaminate the information stream with deceptive tactics, some of these will remain, since one cannot purify a stream back to 100%. This can produce additional information asymmetry.

- dynamics. As with biochemical reactions in bacteria,[47] network connections in people are dynamic, not static. Thus, the connections and unit operations discussed in this chapter are only snapshots of approximate reality, like trying to find the network connections in a forest or an ocean.

As a result of these factors, the Players are in some sense each playing a different game! This needs to be reflected in the PFD, where there are parallel paths in place until an "Enforcer" or "game master" causes them to come together into a final decision.

Games can be solved using PFDs and equilibrium.

We can represent games using Process Flow Diagrams (PFDs), and then solve the various stages with reaction equilibrium concepts. Let's take a look at an example.

Example 4-2. Decision equilibrium for the Battle of the Sexes game.

_____

We will solve the game from Example 1-4. The game is

	b1 = hike	b2 = movie
a1 = hike	*-3.5*, -2.3	*-0.2*, -0.4
a2 = movie	*+0.3*, +0.5	*-2.6*, -3.3

We will take the given Pains as the Δg^0 values, and assume a neutral Enforcer. Assume that all initial values are 0.50 mol. With learning and education, these initial values could change, to bias particular outcomes.

SOLUTION
We draw the PFD for this process. Each way that we draw the PFD represents a different dialogical process, and we might well have control over this design. Changing the order of the unit ops, or allowing for creative solutions in part of the process, greatly changes the outcome. Below is one way to do it, in which the decisions are first made in a parallel process:

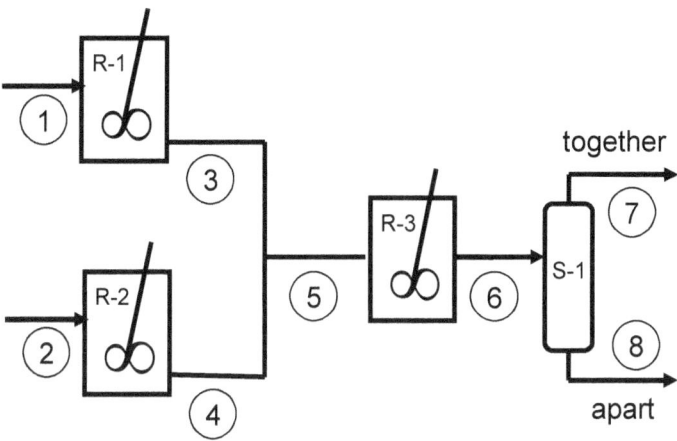

In this process Player A makes a choice in reactor unit R-1, while B makes a choice in R-2. These join together in R-3. After they make their decision, an evaluation occurs in separation unit S-1. If they are together, one decision results. We might add steps after stream 8, depending on whether the two people prefer to give up on the event, or recycle and try again.

We could have chosen to set up the dialogical process as a simultaneous process:

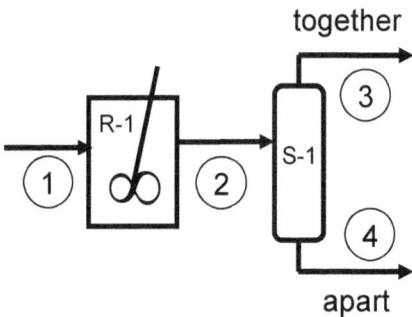

or as a sequential process:

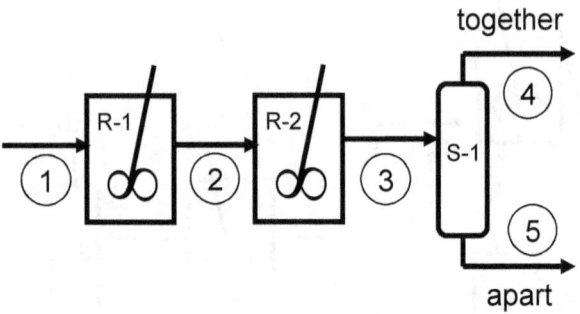

But here we will use the first PFD, for the parallel decision.

The reactions occur in R-1, R-2, and R-3. In R-1 we have reactions 1 to 4:

$$a1 + b1 \xleftrightarrow{\;A\;} A11, \quad \Delta g^0 / RT = -3.5$$

$$a1 + b2 \xleftrightarrow{\;A\;} A12, \quad \Delta g^0 / RT = -0.2$$

$$a2 + b1 \xleftrightarrow{\;A\;} A21, \quad \Delta g^0 / RT = +0.3$$

$$a2 + b2 \xleftrightarrow{\;A\;} A22, \quad \Delta g^0 / RT = -2.6$$

species	initial (n_{0i})	change	equil (n_i)	y (for all)
a1	0.5	$-\varepsilon_1 - \varepsilon_2$	$0.5 - \varepsilon_1 - \varepsilon_2$	$y_i = n_i / \Sigma$
a2	0.5	$-\varepsilon_3 - \varepsilon_4$	$0.5 - \varepsilon_3 - \varepsilon_4$	
b1	0.5	$-\varepsilon_1 - \varepsilon_3$	$0.5 - \varepsilon_1 - \varepsilon_3$	
b2	0.5	$-\varepsilon_2 - \varepsilon_4$	$0.5 - \varepsilon_2 - \varepsilon_4$	
A11	0.0	$+\varepsilon_1$	$+\varepsilon_1$	
A12	0.0	$+\varepsilon_2$	$+\varepsilon_2$	
A21	0.0	$+\varepsilon_3$	$+\varepsilon_3$	
A22	0.0	$+\varepsilon_2$	$+\varepsilon_4$	
			$\Sigma = 2.0 - \varepsilon_1$	
			$- \varepsilon_2 - \varepsilon_3 - \varepsilon_4$	

Solving this we find that
$\varepsilon_1 = 0.3651$
$\varepsilon_2 = 0.0187$
$\varepsilon_3 = 0.0124$
$\varepsilon_4 = 0.3115$

In R-2 we have reactions 5 to 8:

$$a1 + b1 \xleftrightarrow{B} B11, \quad \Delta g^0 / RT = -2.3$$

$$a1 + b2 \xleftrightarrow{B} B12, \quad \Delta g^0 / RT = -0.4$$

$$a2 + b1 \xleftrightarrow{B} B21, \quad \Delta g^0 / RT = +0.5$$

$$a2 + b2 \xleftrightarrow{B} B22, \quad \Delta g^0 / RT = -3.3$$

species	initial (n_{0i})	change	equil (n_i)	y (for all)
a1	0.5	$-\varepsilon_5 - \varepsilon_6$	$0.5 - \varepsilon_5 - \varepsilon_6$	$y_i = n_i / \Sigma$
a2	0.5	$-\varepsilon_7 - \varepsilon_8$	$0.5 - \varepsilon_7 - \varepsilon_8$	
b1	0.5	$-\varepsilon_5 - \varepsilon_7$	$0.5 - \varepsilon_5 - \varepsilon_7$	
b2	0.5	$-\varepsilon_6 - \varepsilon_8$	$0.5 - \varepsilon_6 - \varepsilon_8$	
A11	0.0	$+\varepsilon_5$	$+\varepsilon_5$	
A12	0.0	$+\varepsilon_6$	$+\varepsilon_6$	
A21	0.0	$+\varepsilon_7$	$+\varepsilon_7$	
A22	0.0	$+\varepsilon_8$	$+\varepsilon_8$	
			$\Sigma = 2.0 - \varepsilon_5$ $- \varepsilon_6 - \varepsilon_7 - \varepsilon_8$	

Solving this we find that
$\varepsilon_5 = 0.2856$
$\varepsilon_6 = 0.0263$
$\varepsilon_7 = 0.0127$
$\varepsilon_8 = 0.3496$

Before proceeding to R-3, we needed to solve these reactions and their associated extents of reactions, which were given. Because these are in three separate reactors, we can solve each reactor independently. Now in R-3 we have reactions 9 to 12:

$$A11 + B11 \xleftrightarrow{D} D11, \quad \Delta g^0 / RT = 0.0$$

$$A12 + B12 \xleftrightarrow{D} D12, \quad \Delta g^0 / RT = 0.0$$

$$A21 + B21 \xleftrightarrow{D} D21, \quad \Delta g^0 / RT = 0.0$$

$$A22 + B22 \xleftrightarrow{D} D22, \quad \Delta g^0 / RT = 0.0$$

species	initial (n_{oi})	change	equil (n_i)	y (for all)
A11	0.3651	$-\varepsilon_9$	0.3651 $-\varepsilon_9$	$y_i = n_i / \Sigma$
A12	0.0187	$-\varepsilon_{10}$	0.0187 $-\varepsilon_{10}$	
A21	0.0124	$-\varepsilon_{11}$	0.0124 $-\varepsilon_{11}$	
A22	0.3115	$-\varepsilon_{12}$	0.3115 $-\varepsilon_{12}$	
B11	0.2856	$-\varepsilon_9$	0.2856 $+ \varepsilon_9$	
B12	0.0263	$-\varepsilon_{10}$	0.0263 $+ \varepsilon_{10}$	
B21	0.0127	$-\varepsilon_{11}$	0.0127 $+ \varepsilon_{11}$	
B22	0.3496	$-\varepsilon_{12}$	0.3496 $+ \varepsilon_{12}$	
D11	0	$+\varepsilon_9$	$+\varepsilon_9$	
D12	0	$+\varepsilon_{10}$	$+\varepsilon_{10}$	
D21	0	$+\varepsilon_{11}$	$+\varepsilon_{11}$	
D22	0	$+\varepsilon_{12}$	$+\varepsilon_{12}$	
	1.3818	$-\varepsilon_9 - \varepsilon_{10} - \varepsilon_{11} - \varepsilon_{12}$	$\Sigma = 1.3818$ $-\varepsilon_9 - \varepsilon_{10} - \varepsilon_{11} - \varepsilon_{12}$	

Solving this we find that
$\varepsilon_5 = 0.05600$
$\varepsilon_6 = 0.00037$
$\varepsilon_7 = 0.00012$
$\varepsilon_8 = 0.05824$

NOTES:
- coupling functions. In determining the Pains for the reactions, we used those given. It is important to recognize the coupling that occurs. A choose a1 not only because of the preference for hiking, but because A likes to be with B. This gives an explicit nonlinearity in the game matrix.
- distractions. As is always the case, adding (inert) distractrions into the reaction changes the mole fractions, and therefore changes the equilibrium outcome. Sometimes the change is hard to predict by inspection.
- enzyme. In this solution we have assumed a constant concentration of enzyme. We could adjust for this kinetically if the enzyme concentration were smaller, using Michaelis–Menten kinetics.
- learning or education. If we allow for "learning", we could evaluate the overall ΔG for the reactions, then update the

initial values in a Bayesian manner, to find what initial
values give the lowest final G value.

Although we must solve 12 extents of reactions even for a
simple BoS game, in many ways the equilibrium is easy to solve,
even for this larger systems. Games can be estimated sometimes
intuitively, using Gibbsian methods. After all, solving for a single
equilibrium gives all the final extents of reactions, and therefore the
final amounts, for any reactor unit operation. This also provides a
heuristic advantage, because we can readily see how making single
changes to games will affect the final equilibrium, based on Le
Chatelier's principle.

As the end of the example says, our job has three parts required for
further design:

- For any given initial concentrations ($\mathbf{n}^0$), we need to
  develop the scheme to find the vector ($\boldsymbol{\varepsilon}$) by minimizing
  the function (G). The $\mathbf{n}^0$ therefore represent a type of
  Bayesian prior information, about a person's beliefs
  coming into a game. Thus, the prior bias is separated from
  the pain matrix.

- For a given perspective on the game, the players can
  "learn" a better initial $\mathbf{n}^0$, which lowers the pain.

- The players can alter the PPPP, by introducing new Players
  and Possibilities (i.e., new species), by adjusting the Pains
  and Perspectives (i.e., altered A, B, C and so on), including
  altering the Pains of the final "executive reaction".

Exercises

1. *PFD for cumene.* Use Kirk-Othmer or a similar reference, and draw a PFD for cumene production.

2. *Battle of the Sexes game.* Follow the example given and solve for the BoS game, but use initial values of $n_{a1,0} = 0.70$, $n_{a2,0} = 0.30$, $n_{b1,0} = 0.90$, and $n_{b2,0} = 0.40$. Note that the ratios and the total amounts are different for each Player.

References

Biegler, Lorenz T.; Grossmann, Ignacio E.; Westerberg, Arthur W. *Systematic Methods of Chemical Process Design.* Prentice-Hall (1997).

Turton, Richard; Bailie, Richard C.; Whiting, Wallace B.; Shaeiwitz, Joseph A.; Bhattacharyya, Debangsu. *Analysis, Synthesis, & Design of Chemical Processes*, 4th ed. Prentice-Hall (2012).

Walas, Stanley M. *Chemical Process Equipment: Selection and Design.* Buttersworth (1988).

Walker, William H.; Lewis, Warren K.; McAdams, William H.; Gilliland, Edwin R. *Principles of Chemical Engineering*, 3rd ed. McGraw-Hill (1937). William Walker has Penn State roots, since he was the first Penn State Chemistry graduate in 1890. He went on to MIT in later years.

5. Variance. Risk, uncertainty, and bounded rationality.

So a measurement doesn't have to eliminate uncertainty after all. A mere reduction in uncertainty counts as a measurement and possibly can be worth much more than the cost of the measurement. – Douglas W. Hubbard[48]

Even when this more useful concept of measurement is adopted, some things seem immeasurable because we simply don't know what we mean when we first pose the question. – Douglas W. Hubbard.

Measurements have variance.

In his book *How to Measure Anything*, Douglas Hubbard's definition of "measurement" opened my mind in several key ways:

- anything. Can we measure love? Can we measure tenderness? Can we measure a kiss? Yes, if we are careful in defining what each of these we mean in a given context.
- distribution. A measurement is not simply a number with units. Rather, it is a probability distribution with units. More simply, a measurement can be given as a range of numbers with units. From the perspective of Claude

Shannon, a measurement reduces the information needed to specify a quantity – measurement therefore reduces uncertainty.

- cost. Taking measurements has a cost. We must estimate that the value of using the measurement will be more valuable than the cost of taking the measurement.

Monte Carlo simulations can assess uncertainty.

Simulations enable us to model many things, by running many trials. Let's look at a simple example, for finding the numerical value of $\pi \approx 3.1416$.

Example 1-3. Simulating to find π.

Find the value of π from a simulation.

SOLUTION
We will "throw random darts" to find the fraction of area inside and outside the circle. Say we have a square that extends from (x = 0, y = 0) to (x = 1, y = 1). The area is unity. We can define $\pi / 4$ as the fraction of area that a quarter circle takes up, that is enclosed by this square.   In order to make the estimate, we choose a uniformly distributed random number x from 0 to 1, and a uniform random number y from 0 to 1, and then take some number n of (x, y) pairs. We then evaluate

$$r = \sqrt{x^2 + y^2}$$

If the value is less than or equal to 1, the point is "in the circle". Otherwise it is "in the square, but outside the circle". By finding the "fraction of darts" inside the circle, we find $\pi / 4$.

I have done this calculation in Excel and in Mathematica. In choosing n = 100 points, I got these 5 estimates of the fraction of points inside the quarter circle: 0.82, 0.83, 0.77, 0.76, 0.78. Multiplying by 4 and evaluating the 90% confidence interval gives π = 3.168 ± 0.119. If I take more points, I can reduce the uncertainty. Here is a plot of a typical set of 100 points, within a quarter unit circle.

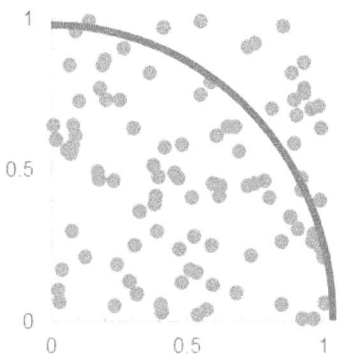

Many simulations can be run on software as simple as Microsoft Excel. As Sam Savage points out, simulations let us use the full nonlinearity of a model, so that we don't fall into the trap of the "flaw of averages",[49] where we consider primarily linear thinking.

We can likewise evaluate uncertainty on a line, a curve, or any complex function, by using simulations. For complex geometries, Monte Carlo techniques are used to estimate volumes.

Tipping points occur at knife-edge decision points.

Why are we concerned with uncertainty, and how to reduce it? In many cases we want to make decisions that give the minimum pain possible (i.e., quite negative!) – for instance, the highest return on investment – but it often happens that as you get a higher average return, you also have a larger risk of harm. By understanding the uncertainties, we can find where we can play to the highest risk level, and thus the highest return level, without causing catastrophe or overwhelming damage.[50]   Furthermore, if you can identify "where the fence" is better than your competitors, you might find that the competition is significantly reduced there, actually reducing the risk from what it otherwise might be.   Thus, calculating risk is a very important exercise.

In Gibbsian games, we have perception functions, Pains, Priors, and other measured values.   Each of these has an associated uncertainty.[2]   Incorporating all of these uncertainties to find the uncertainty of the final strategies is challenging to impossible. However, playing the same Gibbsian game 10 times or 100 times, with the parameters taken properly from the measurement distribution, rather than as a single number, enables us to calculate the values appropriately.   A key revelation is sometimes that the uncertainty in the input measurement distributions renders final predicted strategies of fairly broad distributions; however, we are now able to identify which measurements we need to spend more time with, in order to improve our outcome.

Exercises

1.   *evaluating π.*   Evaluate π, with a known uncertainty, for n = 100, 1,000, and 10,000 points.   Take 10 trials of each as a starting point.   In essence, you are finding the "volume of a 2-dimensional (2D) sphere".

---

[2] Here I use the word "uncertainty" as meaning the estimated statistical variance, not as unknown "black swan" type risks. These are of great importance.

2. *volume of a 4D sphere*. Evaluating the volume of a hypersphere can also be done through simulations.   As given at https://en.wikipedia.org/wiki/Volume_of_an_n-ball, show that the volume is $V = \pi^2 a^4 / 2$

3. *uncertainty for a line*. Say we have the following (x, y) pairs, each with uncertainty. For this problem, we will model the uncertainty as normally distributed, the the one standard deviation numbers given after the ±. Find the best fit of the line, given the uncertainties. Note that the uncertainty of just the (x, y) points given is zero, at y = 3x + 2.5. I used the norminv function in Excel, which you can Google to learn more about. Then I pasted 5 copies of the data below, giving 20 "total data points".

    $(0.0 \pm 0.2, 2.5 \pm 0.3)$
    $(1.0 \pm 0.3, 5.5 \pm 0.1)$
    $(2.0 \pm 0.1, 8.5 \pm 0.4)$
    $(3.0 \pm 0.4, 11.5 \pm 0.2)$

    answer. y = ax + b, where a = 2.59 ± 0.20, b = 2.87 ± 0.40

4. *uncertainty for a game*. Start with the following BoS game, and solve it over and over to calculate the average outcome, as well as the uncertainty in the outcome. The upper number in each block is for Player A, while the lower number is for B. The ± represents the one standard deviation of the distribution, assuming a normal distribution for each.

	b1 = hike	b2 = movie
a1 = hike	*-3.5 ± 0.2*,	*-0.2 ± 0.4*,
	-2.3 ± 0.5	-0.4 ± 0.3
a2 = movie	*+0.3 ± 0.6*,	*-2.6 ± 0.1*,
	+0.5 ± 0.2	-3.3 ± 0.5

References

Hubbard, Douglas W. *How to Measure Anything*. Wiley (2007). This amazing book changed my whole concept of measurement.

Savage, Sam L.; Danziger, Jeff. *The Flaw of Averages: Why We Underestimate Risk in the Face of Uncertainty* (2012). This book provides a nice complement to Hubbard's book on *How to Measure Anything*.

6. Story. Complex people and simplified models.

A mathematical theory ... deals with a
simplified model of the world, a
mathematical model in which only things
pertinent to the behavior under consideration
enter. – John R. Pierce[51]

Stories keep complexities; models simplify reality.

The value of combining stories and models first came to me in a discussion with my colleague Ted Alter in July 2015. The same idea is expressed in the Gintis quote at the beginning of Chapter 1. Models, especially mathematical models, are sometimes thought to be objective, unfettered by normative ideas. Of course, this is frequently false, especially for complex situations.[52] Models by their nature exclude ideas (Pierce quote above). During situations of great information overload, models allow us to wade through the details to attain a solution.

Even in relatively simple physical situations, however, models can lead us astray. For example, in my own technical work into studying quantum van der Waals interactions between atoms, I have found this to be true. Many researchers use a field called "molecular dynamics" to solve for the dynamics and thermodynamics of atomic and molecular systems. Essentially, the researchers know the force field between 2 atoms, and sum all the forces on each atom, solve Newton's law, and step forward in time. Here is the problem: When there are 3 atoms, the force on atom 1 is not simply the sum of 2 atom forces, with 2 acting on 1 and 3 acting on 1. Rather, when 2 acts on 1, atom 3 affects the 2-1 interaction! And our research found that in some cases, not only are 3-body interactions important, but so are 4-body, and higher, even sometimes up to 10-body interactions! By using a simple model, we lose the granularity required to give an accurate solution. And so you say, "Well, that's just a matter of accuracy. Of course a more accurate model gives a more accurate solution." Yes, but how accurate is accurate enough? In the physical sciences, we spend a lot of effort trying to determine what effect is rendered by the next perturbation. Sometimes it is small. And sometimes it is enormous.

In short, I highly recommend learning what you can about the whole story – as much experience as possible – before seeking to model the situation quantitatively. There is often times granularity in the story, that had you seen it, you would try to include in the analysis and design. If you seek to assess a decision-making process in the abstract, without having access to one or more experienced people in that arena, you might well be in for quite a surprise. At the same time, as Gintis suggests, if we know only the story, without some way to tie together the various pieces using a model, we are handicapped. After all, a model contains the accumulation of a great deal of experience. Both story and model have valuable insights to offer.

One final point: It is really hard to define a consequential game! By "consequential", I mean a game in which the game guides one to a decision that can be designed, as opposed to simply having an obvious and trivial solution, either good or bad. Valuable games guide us to new solutions.

There are several ways to piece together a story. An important advantage of having these frameworks for a story, is that 1) we recognize some information that needs to be obtained, and 2) we recognize "surprises", which do not match the information asked for in these frameworks. In short, "if you are looking for nothing, you will find little." In this chapter we will examine several of these perspectives.

Heart of a story: Conflicts, fears, and relationships.

In Chapter 1 I described the multi-P framework for studying strategic decision making processes. In learning the story about a situation, you must identify the basic information of any story:

- characters. Who are the key people, and how might they change? What are their fears? Their trigger points? Who are their friends? Adversaries?

- setting. Where? When? Duration? How large? Geography? Sensory data?

- plot. What are some of the events that have happened, that have changed how people feel about a decision, or themselves, or others? What challenges have occurred that will build courage, or fear, or hope, or sadness?

- climax. What range of climax points are people willing to risk? This gives the potential tolerance for change.

- resolution. What Possibilities, Pains, and Perspectives are in current scenarios?

- details. List any details you hear, read, see, or sense. Keeping a file of these details can be useful later, during pattern recognition.

It is useful to examine some other story frameworks, because by examining the same story in multiple ways, one comes to gain a better sense of how to model a situation. A goal is to lose as little critical information as possible. In the beginning, the nature of this critical information is not always clear.

The Pixar story format **contains 6 "blanks"**.

In his book *To Sell is Human*, Daniel Pink describes the essence of a Pixar story[53]:

Once upon a time, _____.

Every day, _____.

Then one day, _____.

Because of that, _____.

And because of that, _____.

Until finally, _____.

I could write a story about this book:

Once upon a time, Darrell studied colloid science.

Every day, he wrote proposals and papers with students.

Then one day, he co-taught a large MOOC course.

Because of that, he got the courage to formally study his old passion, Physics of Community. He ran into many difficulties and trials.

Because of that, he surrounded himself with experts and colleagues who could help and encourage him with the book.

Until finally, he published the book.

Of course, I could add significant detail to bring this more alive. And I can write such a story about any time period, and include much more detail. Importantly, this story format helps in identifying key markers of a story.

The Made to Stick story format has a useful acronym.

Perhaps the best book I have read about writing and communicating is the book *Made to Stick*, by Chip and Dan Heath. They offer an acronym SUCCES, which guides steps to take in developing a story.

S. Simplicity (essential core of idea). 10 ideas won't stick. 1 might. Be a master of exclusion.

U. Unexpectedness. Violate the reader's expectations!

C. Concreteness. Use sensory information. Avoid abstract thinking. Tiny red larva worms, disgusting vomit smell, two-tone wood grain.

C. Credibility. The idea has to contain within itself a "try before buy" test.

E. Emotions. Clarify the feelings and emotions. We are wired to feel things for individual people, not for abstractions. Fear, sadness, anger, joy, disgust.

S. Story. A narrative is helpful. Hearing stories acts as a mental flight simulator, trying something before doing it.

The Kipling WWWWWH formula requires facts.

I don't actually know how often journalists or police use this little WWWWWH formula. It appeared in *Just So Stories* by Rudyard Kipling in 1902, as a poem near the end of "How the Elephant Got His Trunk" (or "The Elephant's Child"):

> I keep six honest serving-men
> (They taught me all I knew);
> Their names are What and Why and When
> And How and Where and Who.

Who. Characters, connections, friends, adversaries.
What. Issue, controversy, objects, money.
When. Hour, day, season, year, decade, century, duration.
Where. Location, geography, setting, details.
Why. Motives, intentions.

How. Operations, process, steps.

The scientific method story format provides a logical basis.

In my research lab, we use the scientific method for all of our research! The hardest part for most PhD students to learn … is defining an interesting question! My students usually "play" in the lab – that is, try many experiments based on their knowledge, with proper safety measures, until something interesting arises. Once they have something interesting, they perform the scientific method:

- **question**. Define an interesting question, usually answerable by a plot, image, table, or equation in the STEM fields.
- **background**.   Gather   background   from   preliminary experiments, the literature, and other experts.
- **hypotheses**. Make educated guesses, which is a guess of the plot, image, table, or equation. This at least guides the proper axes and format for the result.
- **data**. Clarify the experiments or modeling processes necessary to test the hypotheses, and take the data.
- **results**. Collect the results and analyze them statistically or otherwise.
- **conclusions**. State the outcome of the hypothesis test (i.e., where you are, or "x"), as well as the next steps (i.e., where you are going, or "dx/dt").

The scientific method is iterative and self-intersecting, and so frequently after doing one step, one must return to a previous step to refine it.

Historical methods examine events using internal consistency tests, external consistency tests, witness examination, and other methods. For example, eyewitnesses are usually more credible than hearsay. Other methods, such as bringing to bear scientific or statistical evidence, is also important. The historical method seems more intricate than the scientific method to me, but perhaps that is simply because I am a scientist, not a historian!

Exercises

1. *Arena for strategic decisions.* List 5 arenas where strategic decisions are especially interesting or important to you, and describe why. Rank the 5 arenas, since we will look at the top arena in subsequent problems. Preferably, these areas will be actionable for you, rather than about "global peace" or an international challenge in which we have little say. You might choose problems in your own community, your own workplace or school, your own family, your own street, or similar. For this exercise, you might emphasize games that you are prepared to battle for, rather than choosing battles that are currently far beyond your "zone of proximal development" (Vygotsky's words).

2. *Pixar story.* Write a 5 page summary story of your top-rated arena from Problem 2-1. Use the Pixar model discussed in this chapter. Be sure to include the usual story items like characters, setting, plot (cause and effect leading to change), and conflict. How might you like to see the story end?

3. *SUCCES acronym.* List additional details you can add to the Pixar story, based on the Heath & Heath SUCCES acronym.

4. *Kipling formula.* Apply the WWWWWH formula, and list any additional facts enter your story.

5. *Scientific and historical methods.* List additional points you can add to your story in Problem 2-2, based on what a scientist or historian might ask.

6. *conversion to a game.* Previously we examined classical game theory, including Players, Possibilities, Pains, Perspective, and Prior. Convert your story into one or more games.

7. *PFD.* For the story developed in exercises 7-2 through 7-5, draw a PFD.

8. *possibilities.* Are there additional unit operations, streams, recycles, or other ideas that are suggested by the PFD? State them clearly.

References

Heath, Chip; Heath, Dan. *Made to Stick: Why Some Ideas Survive and Others Die.* Random House (2007). This is the best book I have read about conveying ideas to others.

Pink, Daniel H. *To Sell Is Human: The Surprising Truth About Moving Others.* Riverhead (2013). This book has several gems in it; the Pixar format summary has been very useful to me.

7. Design. Creating and Analyzing Scenarios.

> *[this book] ... may seem altogether too much*
> *like a manual for swindlers. Perhaps I can*
> *justify it in the manner of the retired burglar*
> *whose published reminiscences amounted to*
> *a graduate course in how to pick a lock and*
> *muffle a footfall: The crooks already know*
> *these tricks; honest men must learn them in*
> *self-defense. – Darrell Huff*

Design scenarios can be created, analyzed, heuristicized.

PC solutions aim to give fast, accurate designs for power and collective decisions. We gain ability to finding tipping points and black swans, and to assess risks and probabilities. In game theory this is called mechanistic design.

We can now establish a "design process" for social-political-economic-business-cultural-etc problems. This process follows the engineering method.

- problem. Define the goals of the analysis or design.
- background. "Size up" the Problem by gathering information – some which might be irrelevant in the end – from the story.
- hypothesis. Hypothesize scenarios of the future, for what you want to see (or not) happen.

- approach.   Start using the P's: Identify the <u>Players</u>. Arrange their "unit operations" in <u>Process</u> Flow Diagrams (PFDs). Establish the <u>Possibilities</u> and decision reactions that could occur. Learn about the <u>Prior</u> biases of the Players for the various Possibilities. Using independently-assessed perception functions, establish the <u>Pains</u> (equal negative utilities) for each block in a game.
- results. Calculate the <u>Probabilities</u> for the strategies, being aware of "black swans", "tipping points", and other danger or victory outcomes.
- conclusions.   Provide "translated" stories of outcome scenarios, a summary of outcome distributions, new information needed, and a redefinition of goals.

Arenas. Economics, politics, business, culture.

Here are some problems that we can now study, or which at least have me very interested:

- markets. Supply-demand curves for monetary and non-monetary markets.
- elections. Voting with 3 to n Players, including coalitions and other factors.
- commons.   Study tragedy of the commons issues like fishing, climate change, and others as Elinor Ostrom would have studied.
- internal business cycles. Examine dynamics of marketing v R&D timing problem.

- government. I worked with a student concerning the Pennsylvania budget process in 2015, when Governor Wolf held out for a severance tax in the budget.

- war. Are there new ideas we can learn about mutually-assured destruction (MAD), the role of hunger in tipping off war, and education?

- terrorism. Can we detect patterns that enable terroristic activity, and therefore which can be used to design disabling mechanisms?

- love. What can we learn to help this?!!!

- education. What is the role of learning in making Bayesian iterations faster, with more favorable outcomes for fitness and success? Can I learn to bias my initial concentrations better, or play particular games with improved Perspectives? Will I find that the Wild Scholars techniques could work better?[54]

- friend networks. Can we explain Dunbar's number for size of a friend network? It is about 150 people.

- heuristics. What is the role of Le Chatelier's principle?

- information. Can we develop heuristics that guide our future thinking? As sophistication grows, education might well be more, not less important! It becomes harder to Google entire algorithms of thought.

- matching. Following from the work of Shapley[55] or Roth[56] and others, can we improve matching algorithms? We are currently working on a college selection algorithm for high school students, which we think by-passes many of the shortcomings of existing models.

- ecosystems. Can we model a culture?[57] It is very difficult to model a farm, a forest, a fishery, and other ecosystems. We want to model communities using PC.

Using the techniques of PC, we can study power – for instance, information asymmetry or fear-uncertainty-doubt. By knowing the levers of power quantitatively (Problem, Processes, Players, Possibilities, Priors, Pains, Perspectives), systems can be designed to favor particular outcomes.

We can both borrow from and contribute to the study of the natural sciences. For example, evolutionary design is already used in chemistry (e.g., Frances Arnold of Cal Tech). Can we improve our results in evolutionary game theory?[58] Can we improve genetic algorithms and agent-based modeling?

Many fundamental and applied challenges remain.

Here are some of the first challenges that I hope to solve for Physics of Community.

- temperature and pressure. These must be defined more clearly. Temperature might be expressed as a Weber-Fechner function, or in terms of arousal[59].
- heat. What is "energy in motion", when a temperature gradient is involved?
- kinetics. In this book we have examined primarily reaction equilibria. However, the methods apply to kinetic processes as well, although they are more complicated usually.
- play style. If I am a person with a certain FLAVORS Perspective profile, how should I play games in order to minimize my total Pain? For example, I might be Selfish, but in order to maximize my own take, I might play Overall.

- tipping points and instabilities. What factors lead to oscillatory reactions,[60] tipping points, out of control cycles, and similar phenomena?
- perception functions. Why choose 0 to 4, or 0 to 6, or other, for coefficients and amounts?
- enzymes and catalysts. What does it mean to catalyze reactions accounting for various FLAVORS? Is there anything useful to be gained by examining the EC system and codes?[61]

References and Notes

[1] Gintis, Herbert. *The Bounds of Reason: Game Theory and the Unification of the Behavioral Sciences*, revised Edition (2009).

[2] Winston, Wayne L. *Operations Research: Applications and Algorithms*, 3rd ed. Duxbury (1994). This is a thick and general operations research book that contains the basic methods of linear and nonlinear programming, classic game theory, network theory, and many other topics.

[3] Bronson, Richard; Naadimuthu, Govindasami, *Schaum's Outline of Operations Research* (1997). This book gives a simple and quick overview of the subject.

[4] Pfeffer, Jeffrey. *Power: Why Some People Have It and Others Don't* (2010).

[5] Gaventa, John. *Power and Powerlessness: Quiescence & Rebellion in an Appalachian Valley* (1982). Gaventa discusses three dimensions of power, and in the third, he describes how the elites shape the vary issues, so that the understanding of the issues works to their favor. As a grandson of coal miners, and being raised in West Virginia, this book resonated with me. The description of Tony Boyle's hired hit on Joseph "Jock" Yablonski and some of his family was chilling.

[6] Boulding, Kenneth E. *Three Faces of Power* (1990).

[7] Dahl, Robert A. *Who governs? Democracy and Power in an American City* (1964). This is one of the books I read for Bob DiClerico in 1991.

[8] Alinsky, Saul D. *Rules for Radicals: A Practical Primer for Realistic Radicals* (1972). This book, which is sometimes vilified, at times makes one angry, and at times makes one laugh at loud.

[9] Adams, Frank; Horton, Myles. *Unearthing Seeds of Fire: The Idea of Highlander* (1975). The use of dialogue to solve human problems is extraordinarily powerful.

[10] Horton, Myles; Freire, Paulo. *We Make the Road by Walking: Conversations on Education and Social Change* (1990). This is one of the best books I have read on education, dialogue, and action.

[11] Freire, Paulo; Ramos, Myra Bergman. *Pedagogy of the oppressed* (1970).

[12] von Neumann, John; Morgenstern, Oskar. *Theory of Games and Economic Behavior*. Princeton (1944). The seminal book on game theory.

[13] Spaniel, William. Game Theory 101: The Basics (2011). This is a simple, short, nice introduction to game theory. I always tell my own

students to start with this book. Spaniel has also written a more full text *Game Theory 101*, as well as an interesting book on *War*.

[14] Leyton-Brown, Kevin; Shoham, Yoav. *Essentials of Game Theory: A Concise Multidisciplinary Introduction* (2008). This is another nice introduction to game theory, written at a slightly more sophisticated level than Spaniel's book above. In addition, Leyton-Brown and Shoham teamed up with Matt Jackson to offer a nice MOOC on Game Theory.

[15] Osborne, Martin J.; Rubinstein, Ariel. *A Course in Game Theory* (1994). This book is written at an intermediate level.

[16] Fudenberg, Drew; Tirole, Jean. *Game Theory* (1991). This is the "grown up version" of game theory, written by masters of the craft. Tirole won the Nobel Prize in Economics in 2014.

[17] Ostrom, Elinor. *Governing the Commons: The Evolution of Institutions for Collective Action* (1990). Ostrom won the Nobel Prize in Economics in 2009, and wrote about 8 design principles for Common Pool Resource (CPR) institutions.

[18] Hardin, Garrett. "The Tragedy of the Commons." *Science*, **162**, 1243-1248 (1968).

[19] Jeremy Bentham thought about human decisions being made in terms of pleasure and pain. https://en.wikipedia.org/wiki/Jeremy_Bentham.

[20] von Neumann, John; Morgenstern, Oskar. *Theory of Games and Economic Behavior*. Princeton (1944). They give the result for mixed strategies in Section 18.2.5, for 2 Player, 0 sum games.

[21] See his page xii in his preface.

[22] Herbert Gintis gives a different – and very nicely-worded – definition of "rational" in his books, especially *Bounds of Reason*. See p. 1 of the Revised edition.

[23] Veloz, Tomas; Razeto-Barry, Pablo; Dittrich, Peter; Fajardo, Alejandro. "Reaction networks and evolutionary game theory," *J. Math. Biol.*, **68**, 181-206 (2014). In their abstract they write, "In this 'socio-chemical metaphor' molecular species play the role of agents' decisions and their outcomes, and chemical reactions play the role of interactions among these decisions."

[24] Mirowski, Philip. *More Heat than Light: Economics as Social Physics, Physics as Nature's Economics.* (1991). Mirowski explains that most physics representations have not explained much in the economic sciences.

[25] Dill, Ken A.; Bromberg, Sarina. *Molecular Driving Forces: Statistical Thermodynamics in Chemistry & Biology*, 1st ed. Garland (2002). This is an outstanding book on stat thermos. It is called "introductory", but I find some of the explanations to be extraordinarily insightful.

[26] Pauling, Linus. *The Nature of the Chemical Bond*, 1st ed. Cornell (1939). Pauling understood the mathematics of quantum mechanics deeply, as indicated for instance in his book on *Introduction to Quantum Mechanics*.

[27] See the Wikipedia article on Chemical Reaction Network Theory at https://en.wikipedia.org/wiki/Chemical_reaction_network_theory.

Professor Rutherford Aris of the University of Minnesota, whom I spoke with on several occasions, was one of the originators of this field. Professor Aris was a distinguished and amazing man, who was not only a successful chemical engineer – elected to the National Academy of Engineering – but also a professor of Classics at the University of Minnesota, and a noted expert in paleography (study of ancient and historical handwriting).

[28] The shift catalysts are described in https://en.wikipedia.org/wiki/Water-gas_shift_reaction.

[29] Gintis, Herbert. *The Bounds of Reason: Game Theory and the Unification of the Behavioral Sciences.* Princeton University Press (2009). Gintis' book opens with a very nice description of "rational", in terms of beliefs, preferences, and constraints (BPS model). I love his description.

[30] The relationship between loudness and decibels is not exactly logarithmic. See the description and references at https://en.wikipedia.org/wiki/Decibel and https://en.wikipedia.org/wiki/Loudness.

[31] See https://en.wikipedia.org/wiki/Weber%E2%80%93Fechner_law.

[32] Having finite willpower is sometimes referred to as "ego depletion". See https://en.wikipedia.org/wiki/Ego_depletion and references in this site.

[33] For corruption I think of the Saul Alinsky quote: Life is a corrupting process from the time a child learns to play his mother off against his father in the politics of when to go to bed; he who fears corruption fears life.

[34] Miller, George A. "The Magical Number Seven, Plus or Minus Two: Some Limits on our Capacity for Processing Information," *Psych. Rev.*, **63**, 81-97 (1956).

[35] Simon, Herbert A. "How Big is a Chunk?" *Science*, 183, 482-488 (1974). This is a fun and extraordinary insightful article that addresses Miller's article on the magic number 7.

[36] See https://en.wikipedia.org/wiki/Industrial_catalysts.

[37] See https://en.wikipedia.org/wiki/Jeremy_Bentham for an introduction. I first heard of Bentham from my friend Jack Matson.

[38] This statement in Denbigh about the entropy of mixing being involved with reaction equilibrium was absolutely eye-opening to me, even though I had seen the equations many times before! I saw this returning from my first American Economics Association meeting in Boston, in January 2015. It was where I first realized the role of entropy in decision making, and it was exhilarating!

[39] A brief introduction to thermoreceptors is given at https://en.wikipedia.org/wiki/Thermoreceptor.

[40] Rogers, Everett M. Diffusion of Innovations, 5th ed. (2003).

[41] Cohen, Michael D.; March, James G.; Olsen, Johan P. "A Garbage Can Model of Organizational Choice," Administrative Science Quarterly, **17**, 1 (1972). The view in this book appears to me not much different than a CSTR chemical reactor.

[42] Kingdon, John W. *Agendas, Alternatives, and Public Policies*, Update Edition, with an Epilogue on Health Care (2nd Edition) (Longman, 2010). I read an earlier edition of this book in 1991, with Bob DiClerico. This book uses Cohen et al's "garbage can model".

[43] Oreskes, Naomi; Conway, Erik M. *Merchants of Doubt: How a Handful of Scientists Obscured the Truth on Issues from Tobacco Smoke to Global Warming* (2011). There are several books like this now.

[44] Walker, William H.; Lewis, Warren Kendall; McAdams, William H. *Principles of chemical engineering*, 1st ed. McGraw-Hill (1923). This seminal book was perhaps the first textbook of chemical engineering. William Walker got the first graduate degree in chemistry from Penn State.

[45] Pfeffer, Jeffrey. Power: Why Some People Have It and Others Don't. Harper (2010).

[46] Alinsky, Saul. Rules for Radicals: A Practical Primer for Realistic Radicals. Vintage (1989, original 1971, shortly before Alinsky's death). Parts of this book make me laugh out loud – I think of gaining power with opera attenders. This book brings a strong reaction from most people I know who have read it; even some of my more "radical" friends find this book to be radical!

[47] Feist, Adam M.; Herrgård, Markus J.; Thiele, Ines; Reed, Jennie L.; Palsson, Bernhard Ø. "Reconstruction of biochemical networks in microorganisms." *Nature Reviews: Microbiology*, 7, 129-143 (2009).

[48] Hubbard, Douglas W. *How to Measure Anything*. Wiley (2007). This amazing book changed my whole concept of measurement.

[49] Savage, Sam L.; Danziger, Jeff. *The Flaw of Averages: Why We Underestimate Risk in the Face of Uncertainty* (2012). This book provides a nice complement to Hubbard's book on *How to Measure Anything*.

[50] Bill Joyce is an alumnus of our Chemical Engineering Department at Penn State. He once gave a talk in our department, in which he described how higher risk gives a higher rate of return, and he worked hard to "play at the highest risk" he could, while avoiding irreparable harm. Return on investment is not constant with the amount invested or put on the line.

[51] Pierce, John R. An Introduction to Information Theory. Symbols, Signals and Noise. 2nd ed. Dover (1980, 1961 first edition). This is a nicely-written book, which explains many concepts about information theory clearly.

[52] Miller, Claire Cain. "When Algorithms Discriminate," NY Times, 2015jul09. See the full story at
http://www.nytimes.com/2015/07/10/upshot/when-algorithms-discriminate.html?_r=0, and the original article at
http://www.andrew.cmu.edu/user/danupam/dtd-pets15.pdf.
The authors found that changing the gender setting on Google from male to female resulted in fewer hits for an ad for high paying jobs.

[53] Pink, Daniel H. *To Sell Is Human: The Surprising Truth About Moving Others*. Riverhead (2013).

[54] Velegol, Darrell. *Wild Scholars* (2011). My book on education in k-12 schools and beyond.

[55] Gale, D.; Shapley, L.S. "College Admissions and the Stability of Marriage." *The American Mathematical Monthly*, **69**, 9-15 (1962). Shapley did much of the original work on matching algorithsm, and won the Nobel Prize in Economics in 2012 along with Alvin Roth.

[56] Roth, Alvin E. *Who Gets What — and Why: The New Economics of Matchmaking and Market Design* (2015). Roth describes his work in the "kidney market", in the "education market", and numerous other arenas that we don't always think of as markets.

[57] For modeling cultures, I know of Hofstede's model:
https://en.wikipedia.org/wiki/Hofstede%27s_cultural_dimensions_theory.
But I don't yet know how effective it is.

[58] Background on evolutionary game theory, with references, can be found at http://en.wikipedia.org/wiki/Evolutionary_game_theory.

[59] See https://en.wikipedia.org/wiki/Arousal for information about arousal and an introduction to its parts.

[60] The BZ reaction is described at:
(https://en.wikipedia.org/wiki/Belousov%E2%80%93Zhabotinsky_reactio
n). It is one well-known example of an oscillatory reaction.

[61] The EC system and codes for enzymes is described briefly at
https://en.wikipedia.org/wiki/Enzyme_Commission_number.

Index